Creative
SOLDERED JEWELRY
& ACCESSORIES

Date: 9/23/15

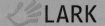

LARK

An Imprint of Sterling Publishing
387 Park Avenue South
New York, NY 10016

Text and photography © 2015 by Lisa Bluhm

ISBN 978-1-4547-0816-2

Distributed in Canada by Sterling Publishing
c/o Canadian Manda Group, 165 Dufferin Street
Toronto, Ontario, Canada M6K 3H6
Distributed in the United Kingdom by GMC Distribution Services
Castle Place, 166 High Street, Lewes, East Sussex, England BN7 1XU
Distributed in Australia by Capricorn Link (Australia) Pty. Ltd.
P.O. Box 704, Windsor, NSW 2756, Australia

For information about custom editions, special sales, and premium and corporate purchases, please contact Sterling Special Sales at 800-805-5489 or specialsales@sterlingpublishing.com.

Every effort has been made to ensure that all the information in this book is accurate. However, due to differing conditions, tools, and individual skills, the publisher cannot be responsible for any injuries, losses, and other damages that may result from the use of the information in this book.

Manufactured in China

10 9 8 7 6 5 4 3 2 1

larkcrafts.com

Creative
SOLDERED JEWELRY
& ACCESSORIES

20+ EARRINGS, NECKLACES, BRACELETS & MORE

Lisa Bluhm

LARK

CONTENTS

INTRODUCTION
6

SOLDERING BASICS

PROJECTS

INTRODUCTION

According to Wikipedia, soldering is a process in which two or more metal items are joined together by melting and flowing a filler metal (solder) into the joint where the two meet, the filler metal having a lower melting point than the piece itself. Sounds simple, right? Well, it really can be, if you have the right tools, materials, and techniques. I wrote my first book, *Simple Soldered Jewelry & Accessories*, to share my passion of creating soldered art, while keeping things simple and easy for the novice. I am excited to expand on the basic skills and share even more new and creative ideas with you in this book.

Since my last book was released, I have been all over the place teaching classes and helping people to achieve soldering success. I also created my own product line of soldering tools and materials, which has sold all over the world! Who knew this fun little hobby would grow and inspire so many to tap into their creativity, and even start their own businesses?

Soldered art has unlimited options and variations. My first book focused solely on soft soldering, which is done with a tin-based solder and a soldering iron. This is still my favorite way to work. In this book, I have not only offered more challenging projects, but I have also included information about hard soldering. I get asked a lot of questions about the differences between the two and how hard soldering is done. This book will address those questions.

Let me be clear: I am not a metalsmith, nor do I claim to be an expert at hard soldering. My writing and technique come from a novice perspective. In fact, the sterling silver projects in this book were my firsts. We're learning this together! I have compiled here the most basic of hard soldering tools and techniques in order to give you a place to start and an idea of how it is done, so you can decide if you want to pursue it and invest more time and money. I encourage you to read everything first to make informed decisions, gather your tools and materials, practice, and then tackle that first project.

Beyond this book is a whole world of metal melting techniques. If you are ready to move on to more complicated metal projects I encourage you to seek out even more information, techniques, and products. I highly recommend taking silversmithing courses from an experienced artist if you want to further your skills. Then again, this book may give you all you ever need to make the projects you want to create.

I don't know about you, but I am usually so excited to get to the creating part of a project that I tend to skim through instructions and just dig right in. But when I make mistakes, and things aren't working right, I go back and reread the information and find out what I should have done in the first place. If you run into problems, don't give up. Go back over the basic techniques and see what step you may have missed or how to better do what you're trying to do.

Be ready to let perfection go . . . these are handmade items and, as such, they will not be perfect. There will be lumps, bumps, scratches, and dents. These imperfections only add character to the pieces and reveal the fact that they were skillfully crafted by hand, not manufactured by a machine. So, don't be afraid to attempt any project here, make mistakes, and add your own creative spin!

Thank you for reading this book. Enjoy the creative journey, and solder your art out!

SOLDERING BASICS

Soldering is easier than you might imagine. It's simply a method of connecting metals using a bonding agent known as solder. The solder must be heated with a small iron or torch in order to fuse the metals. Sure, there's a bit of science involved, but soldering is simple once you get your bearings. Don't be intimidated by the torch!

The projects in this book include both soft soldered and hard soldered pieces. You may wonder what the difference is between soft and hard soldering. Soft soldering uses tin-based solder and a soldering iron. The process allows you to work with materials at a low melting point of around 450°F (232°C). Hard soldering is done with sterling silver solder. Because of the high melting point involved—1145°F (618°C)—hard soldering projects require a butane torch rather than an iron. Keep in mind that the melting point of the solder, whether it's tin based or sterling silver, should be lower than that of the metal you're soldering it to.

Projects that call for sterling silver can easily be made using other types of metals and soft solder tools. One advantage of mastering both types of soldering is that you can create a soft soldered prototype of a piece from inexpensive metals before you make the final product with a torch and sterling silver. Sterling silver mistakes can be costly!

Given the money and time involved, I prefer soft soldering to hard. I can make about five soft soldered projects in the time it takes me to create one hard soldered project. I'm not a jewelry snob. Wearing a piece that's made of tin instead of sterling silver makes no difference to me as long as it looks great. When both types of solder are polished to a high shine, you can barely tell the difference between the materials.

SOFT SOLDERING

The solder used in most soft soldering projects is lead-free and consists of tin, copper, and a small amount of silver. With a soldering iron and the other materials covered on the pages to come, you can use it to create a wide range of items, including jewelry and home décor pieces.

SOFT SOLDERING WORKSPACE

Having a studio is a luxury but not a necessity. A soft soldering workspace is simple to set up—you can easily solder at your kitchen table. However, working in a well-ventilated space is a must. Most standard-sized rooms or even a garage are suitable.

Wherever you work, you'll need access to a sink and safe electrical outlets. You'll also need a sturdy table and good lighting. Try to position your table near a window so that you'll have both natural light and ventilation. Since you'll be using heat, the surface you work on should be fire retardant. My work surface is a large ceramic floor tile—when solder melts on it, it can easily be melted back off. You can also use a glass cutting board or large piece of glass from a picture frame. When I'm working away from my studio, I use a nonstick fiberglass sheet that rolls up easily. It's inexpensive and portable. Although the solder can't melt through the

| Soft soldering workspace. |

| Fiberglass teflon sheet and ceramic floor tile. |

fiberglass, the sheet does get a bit hot, so I place a cutting board underneath it.

I use a high table and a low chair when I solder (and I often stand while working; the height of my table allows me to do so). An adjustable chair is ideal. Find one that can be raised or lowered to allow your forearms or elbows to rest against the table as you work. This fulcrum action will help to steady your hands as you solder.

Position an extra table near your main work area. The additional space will come in handy when it's time to work on the parts of a project that don't require soldering, like assembling a piece.

You'll need to set up a separate work area if you plan to cut glass. A smooth, hard, clean surface is essential. If debris gets under the glass while you're cutting it, the

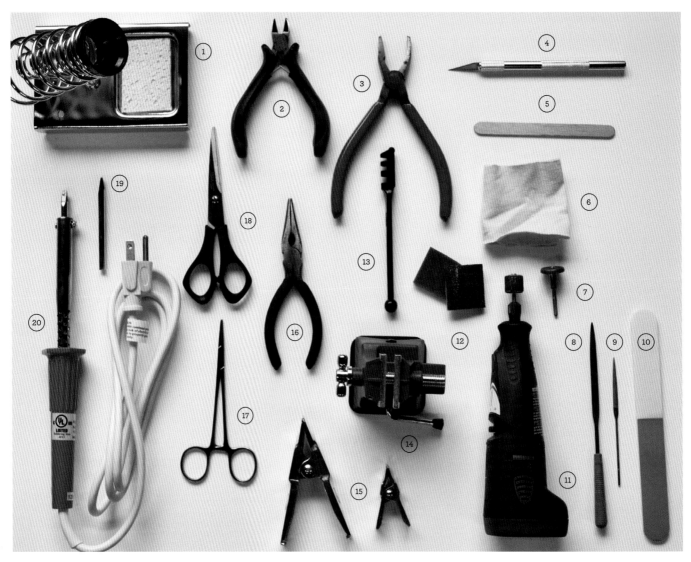

| Soft soldering tools: (1) iron stand with sponge, (2) wire cutters, (3) grozing pliers, (4) craft knife, (5) popsicle stick, (6) polishing cloth, (7) sanding stone, (8) metal sanding file, (9) metal diamond file, (10) fingernail file, (11) rotary tool, (12) sandpaper, (13) glass cutter, (14) vise grip, (15) holding clamps, (16) needle-nose pliers, (17) locking needle-nose hemostat pliers, (18) scissors, (19) iron-clad chisel style soldering tip, (20) 60-watt soldering iron. |

glass may crack. Use a wet disposable towel to wipe up any glass shards or dust—don't use your hands!

Arrange your tools in an orderly fashion so that you can access them easily and safely. You'll likely use your dominant hand to hold the hot soldering iron, so if you're right-handed, keep your heat-related tools on your right side. Put your working tools on your left. This will lessen your chances of grabbing a hot tool unexpectedly.

SOFT SOLDERING TOOLS

A good carpenter never blames his tools. The same goes for a solder artist! Make sure you get the right tools for the job. They don't have to be the most expensive ones on the market. Pricewise, most of the tools I use are mid- to low-range and can easily be found at craft stores, home

BASIC TOOLKIT FOR SOFT SOLDERING

- » COPPER FOIL TAPE ($1/4$-, $5/16$-, AND $1/2$-INCH)
- » CRAFT KNIFE
- » FLUX
- » LEAD-FREE SOLDER
- » SOLDERING IRON
- » IRON STAND
- » CLAMP OR VISE
- » NEEDLE-NOSE OR HEMOSTAT PLIERS
- » FILES
- » SANDPAPER OR A ROTARY TOOL
- » SOFT CLOTH
- » CREAM METAL POLISH

improvement stores, or online. You'll need the items listed in the toolkit (see page 10) for every soft soldering project in this book, so gather them before you begin your work.

Soldering Iron

A soldering iron is your main and most important tool. A soldering iron should not to be confused with a soldering gun. It doesn't have a trigger! A soldering iron conducts heat through its iron tip in order to melt solder.

There are many types of irons on the market. For the decorative soldering we'll do in this book, you want one that's lightweight with an interchangeable tip and the ability to generate even, steady heat at a temperature of around 450–550°F (232–288°C). For easy handling and comfort, the iron should be light and not too large. Various wattages are available. A 60-watt iron works well for me. An iron with a wattage that's too low won't produce enough heat and will slow down the soldering process. On the other hand, an iron with wattage that's too high will melt the solder too quickly. If this is the case, the iron may require a separate thermostat for controlling the temperature.

A brand-new iron can have a strange smell and may even smoke a little when you first heat it up. This is normal and caused by residue from manufacturing. However, if these issues continue, you may have a bad iron.

Always check for cracks, cord wear, and any other type of damage when working with your iron. Your iron should only take a few minutes to heat up, so it doesn't need to sit for long once it's turned on. Turn the iron on when you're ready to solder, and turn it off when you're done. Leaving it on and unattended for too long can be unsafe and ruin the tip or the heater inside. Always use a nonflammable stand that's made specifically to hold a hot soldering iron (see page 12 for more information on stands).

SOLDERING IRON TIP

The tip is the most important part of the iron. It can really make a difference in how your solder flows. The flow refers to how the solder melts and coats the piece you're working on. If the solder flows too quickly, it will set too thin; if it flows too slowly, it can be lumpy and rough. Tips aren't always interchangeable between manufacturers, so be sure to purchase an iron for which replacements tips are available. The best type is a chisel-style tip. It resembles the end of a flat-head screwdriver and allows the solder to flow smoothly. The tip shouldn't be too wide or too thick at the end—⅛ to ¼ inch (0.3 to 0.6 cm) is good. My personal choice is $5/16$ inch (0.8 cm).

Tips for soldering irons are made from plated copper. If the plating is too thin, it will wear off quickly, and the tip will look pitted. Edges may then break off, and holes may form. The best and longest-lasting type of tip I've used has ironclad plating. Finding information on what tips are made of can be challenging, and it may take some trial and error to find one you like. Keep in mind the cost of replacement tips when making your iron selection.

A tip that's been well cared for will allow the solder to melt and flow smoothly, while a dirty or abused tip may not melt the solder at all and can make it seem as though the iron isn't working. Always check to see if the iron is heating up before you discard it. If there's heat coming from the iron but it isn't melting the solder, then the problem is most likely the tip.

New iron tips are often coated in paint, which protects them from oxidation. The paint will burn up and flake off when you heat up the iron, so keep a damp, non-synthetic sponge nearby for wiping off the tip while you work. The key here is to wipe—not press—the iron tip into the sponge, because too much pressure may burn the sponge and in turn contaminate the tip. Keep the sponge damp, as you'll use it continually while you solder.

The first thing to do with a new tip is "tin" it. This seasons the tip for improved soldering and is done by melting some solder on the tip and then wiping it on the sponge. Repeat this action two or three times. Tinning increases heat conductivity between the iron and whatever you're soldering and helps keep the tip from becoming oxidized, which can cause the solder to melt improperly. When you apply the solder, the tip will look shiny. Eventually, though, the solder will become dull and turn black. I suggest that you coat the tip of the iron with solder each time you turn it off. Leave the solder on—don't wipe it off. It will cool and harden on the tip, and when you turn the iron back on, it will melt. Then you can simply wipe it off and start soldering.

Wipe off the tip of the iron on the wet sponge between solder applications. This will ensure that the tip is clean and free of debris. Make it a habit to wipe the tip every two or three times you touch the iron to your project. I usually wipe off the tip each time I put my iron back in its stand.

Tin your tip properly and keep it clean, or it will turn black with oxidation. Once this occurs, it's tough to get the solder to melt. If oxidation happens, you can try re-tinning the tip by melting solder on it and wiping it off several times. If that doesn't work, use a lead-free tip-tinning compound that has solder and flux paste in it. Simply dip your hot iron tip into the compound, wipe it off two or three times, and it should be ready to use. You can also clean a cooled iron tip with a brass-bristled brush, a file, or some other abrasive material that will remove the black oxidation. Sand the tip until you see silver. If you see copper, you've sanded too much—once the silver layer is removed, the exposed tip will begin to disintegrate.

IRON STAND

For safety's sake, you need a place to put your hot iron while you're working. A hot iron can melt or burn anything it comes into contact with. You'll need to rest your iron often while you're working, so invest in a good stand. Choose one with a heavy base that won't tip over. The stand should also have a large coil that sits upright for the iron to slip into and a tray for holding your sponge. The coil will get very hot, so don't touch it while the iron is in it. Clean your sponge often and replace it when it gets worn.

Hand Tools

PLIERS

Pliers are another must-have tool. You'll use them to hold items such as jump rings in place while you're soldering. Needle-nose pliers work well, because they can hold small objects. An even better option, though, is a pair of locking hemostat pliers, the kind used by medical professionals. These pliers are thin and can lock shut, making it easy to hold onto tiny parts and pieces.

GLASS CUTTER

For projects that contain glass, I prefer to use precut material, but when I need a size that can't be purchased, I cut my own. A glass-cutting tool has a tiny steel or carbide blade wheel that scores glass so that it can be broken along predictable, controllable lines. You don't need a specific cutter—any basic one will work. Most cutters have a heavy ball on one end that's used to tap the scored line, which causes the glass to fissure and break apart. (See page 16 for more on glass cutting.)

CLAMP/VISE

Your project will need to be stable and secure while you're soldering it. The piece will get hot, so holding it with your fingers isn't a good idea! You'll need a clamp that won't grip the project too tightly or melt when it gets hot. The clamp should hold the project upright and keep it stable. Spring-loaded metal clamps are a good option. They're sold at most hardware and big-box stores. Use the light-duty type, because heavier clamps can break glass. Always position a clamp on the outside edge of a project—never in the middle.

Another great tool for holding things in place is a vise. You can find vises in all shapes and sizes. Unlike clamps, vises can easily be adjusted for projects of different dimensions. Some have heavy bases, while others clamp onto the edge of a work surface. My favorite vise comes with a 360° swivel that lets me work on a project from many different angles.

You might also consider purchasing a helpful gadget called a jeweler's third hand. This device has moveable arms with an attached alligator clamp. It's great for holding pieces together and will free up your hands.

GROZING PLIERS

Grozing pliers are used to grab glass and break it apart or to break off small jagged edges of glass. They can be used on porcelain, too.

BEADING TOOLS

Some of the projects in this book require beading tools. Be sure to keep them away from your soldering area! The materials used in soldering can corrode the metal on the tools. Some of the items you should have on hand are listed below.

You'll need both light- and heavy-duty *wire cutters* to cut the wire that's used as a base for a project or as part of a design.

Concave *bending pliers* are my favorite tool for shaping wire into curves or creating solder scrolls. They're the best tool for creating the loops on headpins and allow you to make bead charms that are consistent in size and shape.

Flat *needle-nose pliers* will bend wire and hold items in place while you solder.

Round-nose pliers are used to create loops and rounded shapes from wire.

You'll need *crimp pliers* for finishing the ends of beading wire with crimp tubes. Crimp pliers have two notches: one squishes the crimp tube and makes a slight curved indentation; the other notch squeezes the tube, folding it in half.

Finishing Tools

Adding a smooth, textured, or colored finish can accent the overall style of your project. You have lots of options to choose from (see pages 23–24 for more on finishing). Basic sanding files and cloths or an inexpensive rotary tool can be used when you reach this stage.

HAND FILES

Files and sanding tools are great for smoothing out rough edges on glass and solder. Depending on the task at hand, a simple nail file can do the trick, or you may need a heavier metal file. Hand files are inexpensive and simple to use. Use the heavy sanding side to smooth out rough spots or lumps of solder. Use the buffing side to smooth what you just sanded or to bring the piece to a high shine.

ROTARY TOOL

A rotary tool will make finishing work a breeze. Rotary tools vary in price (jewelers use the high-dollar kind), but you can find an inexpensive one at your local hardware or big-box store. Both electric and battery-powered models are available. Choose a lightweight rotary tool that can

easily be held in one hand. The tool uses many different bits and tips for finishing. For example, a sanding drum can smooth out bumps and rough spots. A finishing stone will smooth out solder and bring it to a high shine. A diamond-engraving tip can etch glass.

CLOTHS

You'll need a supply of disposable cloths for cleaning and finishing. Soft old T-shirts work great for this purpose. Use them until they're really dirty and then toss them. Soft paper towels are another option.

SOFT SOLDERING MATERIALS

Once you become comfortable with the soldering process, you'll probably gravitate toward certain products and materials. To help you get started, I put together a list of the basic things you'll need. The items shown below and listed on pages 14 and 15 are safe and have given me great results.

| Soft soldering materials: (1) assorted wire, (2) chain, leather, ribbon, (3) beads, stones, (4) paper and ephemera, (5) metal findings, (6) lead-free came, (7) glass, (8) copper foil sheet, (9) copper foil tapes, (10) lead-free solder, (11) acid-free flux, (12) copper sheet metal, (13) metal embellishments. |

Lead-Free Solder

Lead lends strength and durability to solder, but it isn't safe to handle. So make sure your solder is lead free! If the label on the solder says 60/40, that means it contains 60% tin and 40% lead. Look for a label that says "lead free." This type of solder is composed of tin, copper, and a bit of silver. The silver content shouldn't be higher than about 6%. Any more than that, and you'll need more heat to melt it. Solder comes on a spool and is sold by weight—usually in 1-pound and 1/2-pound sizes (or 250 g and 500 g). Choose a solid-wire variety. The kind with flux in the center won't produce a good finish and can be very messy and smoky.

Flux

Flux is a chemical that's applied right before soldering. Flux removes oxidation from the surface so that solder will stick and flow. The type of flux you use can have a big impact on your results. Some fluxes can be very messy and can leave a sticky residue on the surface. Choose a nontoxic flux that's lightweight and water-soluble. Flux is available in both liquid and paste forms. I prefer liquid because it isn't messy. Liquid flux makes coating projects easy—it provides even coverage and can get into tight spaces and textured areas. I've found that liquid flux is the best way to achieve a smooth flow of melted solder.

Copper Foil Tape

Metal can only be soldered to metal. If the surface you're soldering isn't metal—if, say, it's glass—then you'll need to wrap it in copper foil tape. The solder will stick only to the tape, not to the glass. Copper foil tape is made of thin copper that has a heat-resistant acrylic adhesive on the back. It comes in small and large spools and in large sheets and is available in different widths. Choose a size that will cover the edges of your project with room left over at the edges for folding and burnishing (this will create a casing for the item you're soldering). The tape can be cut into different lengths and shapes with scissors or trimmed with a craft knife after it's applied to an object.

Silvered copper foil tape is coated with a layer of tin on both sides, which gives it a nice silver finish. It also makes for smooth soldering. Copper foil is silvered on the front and back, so you don't have to tin the foil first. Silver-backed foils are used when you can see through the object that you're wrapping—the silver will show where the tape is visible. Silver-backed tape only has silver on the adhesive side.

Black-backed tape is another option for the projects in this book. Decorative-edge foils are another possibility. These foils add a nice touch to soldered projects. You can cut your own designs from it or create your own decorative edges.

Decorative tape can be used on any of the projects in this book, depending on what you want your edges to look like. I use it on Christmas ornaments and around the bezels of stones—it acts like crown molding and holds stones in place.

Lead-Free Came

Came is a slender, grooved piece of metal that can be used instead of copper foil. A piece of glass will fit inside the groove of the came, and it can be formed to fit around the glass. This makes quick work of projects in which soldering around the perimeter of the glass isn't necessary. The only spot you'll actually solder is the joint where the ends of the came meet.

Came can impart a bulky, heavy feel to a project, so it's usually only used in decorative pieces. Lead-free came will melt at the same temperature as lead-free solder, so take care not to touch the hot iron tip directly to it. Instead, melt the lead-free solder and let it drop onto the joint, filling it in.

Glass

Clear, thin glass (no thinner than 1 mm) is ideal for the projects in this book that call for glass. Decorative precut glass is easy to find, or you can cut your own. Old picture frames are a great source for clear glass. You can also buy inexpensive frames and use the glass that's in them. Microscope slides are also handy for jewelry projects.

Extras

JEWELRY FINDINGS
There are so many findings to choose from, and they make great finishing touches for soldered projects. You can find pre-fabricated jewelry components such as bezels, chains, and jump rings in most craft stores. I often use jump rings as bails to hold embellishments and to attach other findings.

Many jewelry findings are plated, and this plating can burn off while you're soldering to reveal metal of a different color. To prevent this from happening, choose rhodium-plated findings. Rhodium plating is strong and will usually withstand solder heat. Other options include base metal, nickel, copper, brass, sterling silver, and gold. Pewter melts at the same temperature as soft solder, so don't use it in your projects.

WIRE

Wire can be shaped into bails, used to create complete projects, and made into decorative elements. Choose wire that's made of brass, copper, or nickel. Plated wire works fine unless it's coated with colored or aluminum plating. This plating will prevent the solder from melting onto the wire. Various gauges of wire are available. A lighter-gauge wire is ideal for creating a decorative embellishment, while a heavier wire should be used to make a bezel or a base for soldering (like a bracelet cuff).

METAL EMBELLISHMENTS

Explore the jewelry, paper crafts, and woodworking aisles in your local craft store, and you'll find all kinds of interesting metal items that can be used in your solder projects: stamps, hinges, beads, box corners, charms, decorative paper clips and brads: the list goes on. These items will bring a bit of whimsy and a personal touch to your pieces.

BEADS AND STONES

Large beads and stones are often used as focal points in jewelry projects. With soldering, though, you're not limited to creating your project around pre-drilled holes—you can add attachments wherever you like. With the exception of items made of plastic, most beads and stones can withstand the heat of soldering, although some stones may crack.

OTHER BEADING MATERIALS

In addition to beading wire and jump rings, you'll also need the following items: crimp tubes, head pins, eye pins, chain, leather cord, coil ends, and ribbon.

CRAFT MATERIALS

Some of the mixed-media projects in this book feature paper or decorative items sandwiched between sheets of clear glass. To make these projects, you should use items like photos or scrapbook papers—materials that are thin enough to be sandwiched. Bulky items can cause glass to crack while you're working with it. However, if you want to use a bulky or very valuable piece in this kind of project, try scanning and printing the item. When printing images, use a non-glossy paper and a laser printer or photocopier. The ink from these machines won't run if the piece gets damp underneath the glass.

Here's a list of craft items you can sandwich in between sheets of glass:

SCRAPBOOK PAPERS	COLLAGES
PHOTOS	DRIED OR PAPER FLOWERS
FABRIC	STICKERS
RIBBONS AND TRIMS	FIBER
EPHEMERA	

Finishing Materials

POLISH

Once the solder is cleaned and smoothed out, you can bring it to a high shine using a cream metal polish. Apply the polish with a cotton swab or cloth, and then buff it off with a soft cloth. Over time, the solder will tarnish due to the elements. To restore the shine, just apply the polish and buff the piece again.

SAFETY: *Be sure to read the manufacturer's instructions and the warnings that appear on your tools and materials.*

Glass Cutting

Some of the projects in this book call for glass pieces of various shapes and sizes. I prefer to use pre-cut glass. Once in a while, I need to cut my own, so I've found that glass cutting is a good skill to have. As with most things, a little practice makes all the difference. Try cutting scrap pieces of glass first to get a feel for the work.

SCORING AND BREAKING THE GLASS

1 Trace a line on the glass using a permanent marker. *(photo a)*

2 Hold the glass cutter in your hand as you would a pen, perpendicular to the glass. Starting slightly inside the edge of the glass, push the cutter away from you and use light pressure to score the surface. You may hear or feel a scraping on the glass. *(photo b)*

3 Stop just short of the edge of the glass at the other end. Take care not to roll the cutter's wheel off the other edge, as this can cause a crack you don't want. You'll see a score line that looks like a scratch. *(photo c)*

4 Tap the glass along the scored line with the end of the glass cutter to fissure the score. *(photo d)*

5 Place the glass with the score line over the edge of a table and push the outer piece down. The glass should snap apart on the score line. *(photo e)*

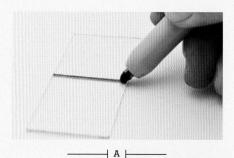

—| A |—

—| B |—

—| C |—

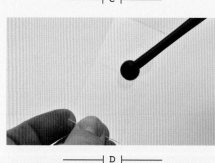

—| D |—

—| E |—

CUTTING CURVES

Curved cuts require a little more work. You can't simply snap the glass apart on the score line. Break lines should be made along the edges of the glass to the curved score.

1 Trace and score your pattern onto the glass as described above. *(photo a)*

2 Score several break-lines from the outer edges of the glass to the edges of the scored pattern.

3 Tap the glass to fracture the break lines.

4 Remove glass from outer scored sections piece by piece. *(photos b–c)*

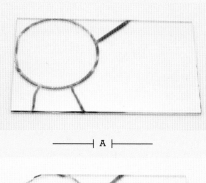

—| A |—

—| B |—

—| C |—

FIXING UNEVEN EDGES

If you didn't score the glass evenly or deeply enough, the edges may be jagged. You can remove these jags with a pair of grozing pliers or some other type of heavy pliers. Grasp a small section of glass near the scoreline with the pliers and pry it downward. Repeat with each jagged area. Then smooth out the rough edges with a file or a sanding drum. Make sure to wear safety glasses when sanding.

Copper Foiling

Copper foil tape serves as a metal base in soldered glass projects. It's pliable and thin, with a heat-resistant acrylic adhesive on one side. It must be applied to the glass before you begin soldering. See page 14 for more information on copper foil tape.

ASSEMBLING SANDWICHED GLASS PROJECTS

Items placed between pieces of glass should be as thin and flat as possible. When sandwiching a thick or uneven item between sheets of glass, make sure it stays level, so that the glass doesn't rock over the top of it. Trim all paper and ephemera so that it fits between the glass and doesn't extend past the edges. Carefully align the glass and the item inside it. Items that stick out past the edges can make the copper foil uneven and become discolored from the soldering heat.

APPLYING THE FOIL

When you've cut your glass and assembled the project, you can apply the copper foil. You want to start with a good foundation, so take care in this first step—it's very easy once you get the feel of it. I often foil several projects while watching TV. The process isn't messy or mentally demanding, and it requires very little room.

1 Make sure the foil is long enough to cover the outside edges of your project and wide enough to fold over the front and back. This will create a frame that holds the project together. Peel the end of the tape away from the roll, but don't cut it yet.

2 Peel away the foil's paper backing and place the end of the foil on one edge of the project. (I like to start my foil on the center bottom edge. I never start in a corner because the bulkiness of the overlapping foil makes it difficult to smooth down.) Center the glass in the middle of the tape so there are equal amounts extending on the front and back sides to be folded over in the next steps. *(photo a)*

3 Rotate the glass in your hand as you wrap the foil around the edges. As you press the tape to the glass, make sure it's even on both sides. Don't fold the edges of the tape over the glass yet.

4 Wrap the foil around the project until you end up where you started. Overlap the end of the foil with the starting point by about ¼ inch (0.6 cm). Then cut the end of the foil to remove it from the roll. You can cut it with scissors or tear it off the roll. I use scissors because they give me a smooth end. *(photo b)*

——| A |——

——| B |——

FOLDING DOWN THE FOIL

What you do from here will vary according to the shape of the project.

Glass-based project with corners and straight edges:

1 Fold the foil in with your fingers on two opposite sides on the glass. Start in the middle, and work your way out to the corners. Neatly and smoothly burnish the foil to the glass. I often use my fingernail to burnish foil. A popsicle stick or bone folder will also do the trick. *(photo a)*

2 Fold the remaining edges over the glass, taking care to make the corners smooth and neat. The goal is to create mitered corners, sealing the foil around the glass. *(photo b)*

——| A |——

|— B —|

|— D —|

|— C —|

|— E —|

with stones or shaped objects make sure there is a folded edge that wraps around the front and back sides so that it grips the item. Without that edge the adhesion is the only thing holding the item in place, and over time that item could slip out. Once a project has been wrapped in foil, it can remain that way for as long as necessary until you're ready to solder it.

—| FOIL TROUBLESHOOTING |—

Tears, bumps, or loose spots in the foil can affect the appearance and strength of the finished project. Luckily, copper foil is easy to remove if you make a mistake. Simply use a craft knife to loosen it and peel it off.

If the foil tears while you're working, remove it and start with a fresh piece. It's possible to patch foil, but that can sometimes be more trouble than it's worth. The patch may lift and move while you're soldering, creating more work and frustration.

If a fold causes a void or divot in the foil, it can't be filled in with solder because there is no metal in the gap for the solder to melt to. Try pulling the fold out carefully (before burnishing) by loosening and unfolding it with a craft knife.

3 Burnish the foil in place by smoothing out any ripples and bumps and squeezing out any air trapped under the foil tape. (*photo c*)

Curved edges on glass and other irregularly shaped objects:

1 Evenly crimp the foil down around the edges of the glass a little bit at a time, working your way around the entire project. There will be many little creases in the foil. Keep it as even as possible. The foiled project should resemble a foil pie tin. (*photo d*)

2 Burnish the foil in place by smoothing it down from the outside edges, pulling in toward the center. Use a firm touch to make the little creases as flat as possible. (*photo e*)

FINISHING UP

Trim the foil to straighten up any crooked edges and to make the foil overlap a little narrower if necessary. Don't trim too much. The ideal

amount of overlap is a minimum of ⅛ inch (3.2 mm). Use a craft knife to cut directly through the foil on the glass or stones. Glass won't scratch, but some stones are soft and scratch easily. Test a spot on the stone before you wrap it to see if the craft knife will scratch it. Once you've made the necessary cuts, peel and remove any excess foil. If some of the adhesive residue remains, clean it off later when you're done soldering.

WRAPPING STONES OR OTHER OBJECTS

The same rules apply for other objects with straight or curved edges. Follow the same steps as above, using the width of tape that best fits the project you're working on. When working

Let's take a moment to review safety tips before you begin soldering. Common sense rules when it comes to working with heat, tools, metal, and glass. Always read labels and follow the manufacturer's instructions when you use a product. Not all products require the same precautions, so be informed. If the product itself has no helpful information, research it online or contact the manufacturer for a Material Safety Data Sheet before you use it.

The following safety rules apply to both soft and hard soldering.

- **CLOTHING AND HAIR.** Wear clothing that's made of a non-synthetic material like cotton. Secure loose, hanging sleeves. Hot solder can drip, so wear an apron or place a towel on your lap to protect your clothes and legs. Remove all jewelry. If you have long hair, pull it back. Don't use flammable cosmetics like hairspray.

- **SAFETY GLASSES.** Yeah, they look funny, but safety glasses can save your eyes! Make it a habit to wear these protective glasses during every phase of a project, whether you're cutting glass, soldering, or sanding.

- **DUST MASK.** A dust mask will keep you from inhaling the dust particles that result from sanding. If you want protection from vapors, use a special mask with a fume trap. Keep in mind that if you're in a well-ventilated room, fumes shouldn't be an issue.

- **GLOVES.** When using patinas and other finishing materials that involve chemicals, you'll need to protect your hands. Rubber gloves will do the trick.

- **WORKSTATION.** Use this checklist to ensure that the space you're working in is safe: (1) If you're working in an area where you eat, keep all chemicals away from food items. Clean the area thoroughly when you're done to prevent contamination. (2) Make sure there's a water source nearby, and keep a fire extinguisher on hand. (3) Keep all flammable objects away from your workstation. Butane canisters should be stored in a cool area away from your torch flame. (4) Allow your tools to cool completely before putting them away. (5) Tape down any loose or hanging cords that could get pulled accidentally.

- **CHILDREN AND PETS.** Soldering requires concentration and focus, so it's best not to do it if there are children or pets around. Keep all of your chemicals, tools, and materials out of their reach. You don't want the kiddos to cut themselves or get burned. Make sure your pooch doesn't chew on an electrical cord or eat something harmful.

Soft Soldering

Melting solder for the first time is very exciting. Of course, reading about it and actually doing it are two very different things. Your first attempt at soldering should be a practice session. Solder again and again until you feel comfortable with the process. You'll soon get a feel for holding the tools and will become familiar with how the solder flows and how much solder you should use.

POSITIONING THE PROJECT

Place your project in a clamp or a vise so that your hands are free for soldering. Position the project so that it's upright, secure, and stable. It shouldn't move while you're soldering. If you use a clamp, make sure to attach it to the outermost edge of the glass. Don't clamp it in the middle—this can put too much pressure on the glass and cause it to break.

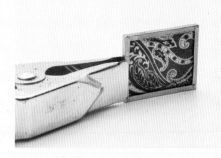

IRON POSITION

Hold the soldering iron in your dominant hand in a pencil grip. Hold it close to the tip end on the handle. It should be at a 45° angle to the edge of the project, with the flat sides of the tip horizontal to the edge. This will allow the solder to flow along the top and over the front and back edges of the project.

APPLYING FLUX

As you may recall, flux prevents the copper foil from oxidizing so that the solder will stick and flow. Start by painting a thin coat of flux on the edge of the copper foil. Applying too much may cause the solder to spit and pop. You don't want the flux to seep under the foil, especially if there's paper between the pieces of glass that could get moist.

Flux only the surface you're working on at the moment (i.e., the one you're about to solder). Once you've added flux to the copper foil, you must solder. If you wait to solder, the foil will patina and become difficult to solder unless you clean it off. If the solder isn't flowing, add more flux. You can continually add flux while soldering to make sure the solder flows freely.

TINNING

Tinning is the process of melting a thin layer of solder over all of the exposed copper foil. Unlike tinning a new solder tip, this process produces a thin base layer. It's a great way to become familiar with the soldering process. Some solder artists start all of their projects with tinning and then go back in and add a bead of solder (a smooth, thick, rounded edge). Once you're proficient at soldering, you may decide to skip tinning and go straight to making a beaded edge with the first pass. In the meantime, here's how to tin:

1 Place the tip of the solder iron on the left end of the project, just in from the corner. If you begin right on the corner, the solder may run down the other side. *(photo a)*

2 Press the end of the solder onto the top flat side of the tip. It should start to melt and flow down the tip. *(photo b)*

3 Once a small amount of solder is on the foil edge, pull the wire away from the tip so that you're not adding too much.

4 Use the iron tip in a painting motion that is one long stroke to draw the solder across, lightly touching the tip to the foil and covering it with solder. *(photo c)*

5 Spread the solder over each edge of exposed foil in a thin coating. Take care not to press the iron tip too hard or hold it too long in one place, as you can actually burn a hole through the foil. *(photo d)*

⊢ A ⊢

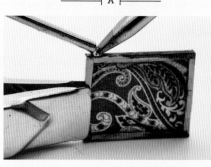

⊢ B ⊢

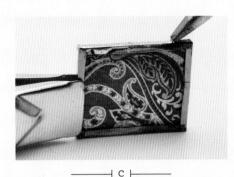

| C |

| D |

MAKING A BEADED EDGE

A beaded edge refers to a solder line that has a nice plump, rounded appearance. This is the desired effect for each soldered project. Creating this type of edge takes some practice, but it isn't necessary to master it in order to create the projects in this book. The appearance of your soldered edges—thick or lumpy, smooth or rough—is a matter of taste. It's your choice. The edges do need to be thick enough for structural strength, and they shouldn't have any sharp areas.

1 To create a beaded edge, position the tip as you did when you were tinning, on the top left edge of the project, just in from the corner.

2 Touch the tip lightly to the foil and press the solder onto the tip to start the flow.

SAFETY: *Be careful when you turn the project and reposition it. The solder, glass, foil, and clamp can be hot to the touch.*

3 The solder should melt and flow down the tip onto the foil. Once it reaches the foil, lift the tip slightly and let it hover just above the foil. There should be a bridge of melted solder connecting the tip to the foil. *(photo a)*

4 Move the soldering iron along the top edge, still hovering above the foil. Maintain a constant bridge of solder by adding more of it from the spool as you go. Your hand will move in tandem. To control the amount of solder, touch the solder to the tip for more and pull it away for less. Beginners tend to add a lot of solder, but it's much easier to work with less. You can go back and add more as needed. Also, if you get too much solder on the project, you can re-melt it, and then hold the project at an angle so that it falls off.

5 If the solder stops flowing, add more flux. *(photo b)*

| A |

| B |

6 When you get to the end of an edge, lift the tip straight up off the edge, separating the tip from the solder and preventing it from running down the side edge.

TIP: *Check the front and back sides of the foil to see if the solder has covered them. If not, add more solder with a couple of extra passes with the soldering iron. There may be enough solder on the top edge so that you don't need to add more. Instead, you may be able to re-melt what's already there, and by lowering the angle of the tip, guide the solder so that it flows down over the foil. Try to solder from one end of the project to the other in a slow, steady, sweeping motion. Keep it even and smooth. Don't stop and start in different spots along the edge.*

7 Smooth and blend bumps or uneven areas by adding more flux and then re-melting the solder, allowing it to flow into a smooth line. *(photo c)*

| C |

8 Once the first side is set and firm, unclamp the project and turn it so you can solder the next side. Continue until all sides are done. Remember: the project is hot!

SOLDERING CURVED SHAPES

Soldering on a curved edge requires a little more maneuvering. You can still use a clamp or vise, but you'll need to work in small increments and repeatedly move the piece around in the clamp. I often hold a curved project up with a clamp, adjusting it as I work so that the solder flows the way I want it to.

SAFETY: *Never put your hand underneath a project while soldering, as the hot solder could drip onto it.*

ADDING JUMP RINGS

To attach a jump ring to your project, you'll need to add some solder and melt it while simultaneously pressing the jump ring into it. There are many ways to do this, but the technique that works best for me is explained below. Just make sure the jump ring has plenty of solder to sink into, making a strong bond between the jump ring and the project.

1 Place a small mound of solder on the spot where you want to attach the jump ring by placing the hot tip near the edge and pressing the solder into the tip until the solder melts. This is one time that you shouldn't add flux first, because you want the solder to form a lump instead of flow. There should be enough flux residue on the solder that it will stick in place but not flow. If the mound of solder won't stick, add a minimal amount of flux. Lift the tip off the solder edge, leaving a mound of solder in place. If there's a point on the mound, lightly flatten it with the tip of the iron. *(photo a)*

| A |

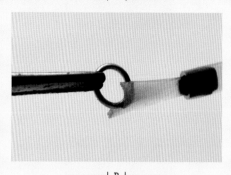

| B |

| C |

2 Paint flux onto the jump ring where you want the solder to melt. Don't skip this step. You always have to flux metal before it will receive solder. If the solder isn't sticking to the metal, add more flux or add more heat from the iron. *(photo b)*

3 Hold the jump ring with a pair of needle-nose or locking hemostat pliers, and place the fluxed jump ring onto the hard mound of solder, open-side down. With your other hand, bring the hot tip of the iron to the mound of solder.

4 When the solder begins to melt, push the jump ring into the melted mound. Pull the iron away once you feel the jump ring nudging up to the foiled and soldered edge. The solder should flow through and over the bottom edge of the jump ring. Hold it still until the solder hardens. *(photo c)*

CREATING DECORATIVE SOLDER BALLS

Little decorative solder balls added to soldered edges can enhance the look of your project. Decorative solder balls are created in the same way the mound of solder used for attaching jump rings are made. Place the tip of the iron near the edge of the project and press the solder into the tip until it melts. Don't add flux first—let the solder form a lump instead of flow. Lift the tip of the iron off the edge of the project, before the solder flows, leaving a ball of solder in place.

The more solder you use, the larger the ball will be. If there's a point on the ball or if it doesn't look smooth when you pull the tip of the iron away, paint a small amount of flux on top and re-melt the solder until it's smooth. Don't hold the iron in place for too long, or the ball will melt and flatten.

TACK-SOLDERING

Decorative wires and embellishments can add a fun and interesting dimension to your soldered edges. The easiest way to add an embellishment is by tacking it onto the edge of your project, meaning that you solder a few key points to secure it. You don't have to put solder all over the embellishment in order to secure it. If you use too much, it can flow all over the element you're adding, which can hide the details of that element.

——┤ A ├——

——┤ B ├——

——┤ C ├——

melt it until a small amount flows and makes a bond. If there isn't enough solder, add a little more. *(photo c)*

PLATING

Many different metal embellishments are available that can be soldered onto a project for an extra bit of detail. If the item you want to use isn't silver colored, you can easily plate it so that it is. Some items may already have plating on them that will burn away, revealing a brass or copper color underneath—you can re-plate them, too. The heat of the iron can cause metals to turn interesting colors. (Brass turns a rosy color, for example.) Heat can also cause the existing plating to burn off in an interesting pattern. If I'm purposely trying to achieve this kind of look, I'll place a soldering iron (making sure the tip is clean so that no solder gets on the piece) onto the metal and hold it there until I get the look I want. To re-plate a metal embellishment, here's what you do.

1 Position the embellishment in the desired area on the project and secure it with a clamp or vise. *(photo a)*

2 Choose a few key points where the embellishment meets up with the soldered edge. This is where you'll tack-solder.

3 Paint some flux on the embellishment and on the edge of the solder. *(photo b)*

4 Put the iron tip on the soldered edge. If enough solder is on the edge of the project to secure the embellishment to the project,

1 Coat the metal with flux.

2 Melt a small amount of solder onto the iron tip.

3 Use the tip to paint the solder onto the metal surface. Use a thin amount of solder—if it's too thick, it'll hide the decorative dimensions.

Finishing

When you're finished soldering, wipe off any excess flux residue with a damp cloth. If flux is left on the solder, it'll cause the silver finish to become discolored. This is the only mandatory step in the finishing process. The other finishing techniques covered here have to do with enhancing the look of your completed project. Different finishing techniques have different effects.

OTHER FINISHES. Paint and ink can also be used as finishes. Some types won't stick to solder, though, so you'll need to experiment. To get the best results, give the solder some "tooth" by sanding and roughing it up a bit. This will make it more receptive to the finish. Paint and ink finishes are best used on decorative items rather than on jewelry.

CLEANING

If you use water-soluble flux as I recommended earlier, then cleaning will be a snap. A project consisting of paper under glass can be cleaned with a water-dampened cloth. Simply wipe off any flux residue. If the project is a solid piece without any paper, rinse the project off with water and pat it dry. If you used another type of flux, you may need a special remover. Check out the product's packaging for information.

If there are black spots or discolored areas on your project, remove them by sanding or buffing the piece. Stubborn sticky spots can be caused by adhesive that may have heated up and seeped out from under the foil. Use a craft knife to scrape these sticky spots off of the solder and the glass.

SANDING AND BUFFING

If you're happy with your solder the way it is, you can skip the sanding process. But if you're going for a brushed or textured effect, you should sand the solder until you like the way it looks.

Sand any jagged edges, rough spots, or unwanted lumps so that the solder appears uniform and smooth. This will make wearable pieces much more comfortable.

HAND SANDING: The simplest way to sand is by hand. You can use files, sandpaper, steel wool, or even a brass-bristled brush. Start out with a heavy grit and work your way up to a finer one. Make sure you're sanding only the solder and not scratching the glass. Once you're proficient at soldering, you'll likely find that the only finishing your projects require is a light touch with a file.

ROTARY TOOL: Sanding is quick and easy if you do it with a rotary tool. This device uses several different bits and tips (see page 13 for more information on rotary tools).My favorite tip is a finishing stone impregnated with emery abrasive. It creates a very smooth, buffed surface. When using a rotary tool, make sure that you don't sand off too much solder and expose the copper underneath. If this happens, apply more solder.

POLISHING

A metal polishing paste will give your piece a shiny finish and protect the solder. Applying it is optional, but it will definitely make the finished product look good! Apply it with a cotton swab and wipe it away with a soft cloth, buffing your piece to a high shine as you work.

PATINAS

Patinas are chemicals that change the color of solder. They're best applied with a bristled paintbrush. Make sure the patina goes only on the solder—if it gets on the glass, it may etch it.

BLACK PATINA. This is the easiest patina to work with on lead-free solder. You can apply as many coats as you need to in order to achieve the degree of color you want. You can also sand off areas of the metal to reveal the silver color again, which can give dimension and depth to your project. Make sure you rinse the project with water after you apply the patina.

COPPER PATINA. This patina is more challenging to work with on lead-free solder. Several layers must be applied to achieve a thoroughly coppered look. Again, make sure you rinse the project with water after you apply the patina.

SAFETY: *Patinas are made with very strong chemicals, so wear gloves and safety glasses when you use them. Be sure to read the manufacturer's instructions and take note of any precautions before you work with them.*

PROBLEM	CAUSES	SOLUTIONS
MY IRON WON'T MELT THE SOLDER.	1. Iron wattage too low. 2. Tip is dirty. 3. Tip is loose.	1. Make sure your iron will get hot enough to melt the solder. It should heat up to a temperature of 450–550°F (232–288°C) and be at least 60 watts. 2. Make sure the tip of your iron is clean. If it's blackened, it may be oxidized or have baked-on contaminants. Clean the tip by sanding and re-tinning it. 3. Try tightening the tip of the iron.
THE SOLDER IS RUNNY AND FLOWS TOO MUCH.	The iron may be too hot.	A high-wattage iron may require a thermostat that will allow you to control its temperature.
THE SOLDER WON'T FLOW. IT STOPS AND LEAVES GAPS.	1. Not enough flux. 2. Not enough continuous heat. 3. Iron isn't hot enough.	1. Make sure to cover the entire area you want to solder with flux. 2. Keep the iron tip in contact with the solder so that it stays molten and flowing. 3. Make sure your iron is set to the right temperature.
THE SOLDER IS LUMPY AND ROUGH.	1. Too much solder. 2. Not enough flux. 3. Not enough continuous heat. 4. Iron too cool.	1. Use less solder. You can add more later if necessary. 2. Use more flux and re-flux the lumps and bumps before re-melting them and smoothing them out. 3. Maintain a bridge of solder with continuous heat. Don't lift the tip off with short strokes. Use a long painting motion to keep the solder molten. 4. Make sure your iron is set to the right temperature.
THE SOLDER RUNS DOWN THE SIDE, LEAVING A LUMP.	1. You're pulling the solder off the end. 2. Too much solder.	1. Stop just short of the very end of an edge and let the solder pool a little before breaking the bridge of solder with the tip. 2. Use less solder. 3. Turn the project, apply flux, and re-melt the lump to smooth it out.
SOLDER SEAMS ARE FLAT, NOT BEADED.	1. Not enough solder. 2. Iron too hot. 3. Iron tip resting on surface. 4. Moving too fast.	1. Use more solder. 2. Make sure your iron isn't too hot. 3. Lift the iron tip up so that it hovers just above the surface and creates a bridge of solder. 4. Slow down and allow the solder to bead up.
SOLDER WON'T STICK.	1. No flux. 2. Wrong type of material.	1. Make sure you coat all metals with flux before soldering. 2. Solder will not melt to aluminum, titanium, and other types of mixed metals.
THERE ARE BLACK SPOTS ON MY SOLDER.	1. Contaminants in flux or solder. 2. Adhesive leaked out.	1. Make sure flux and solder are clean. Black spots can usually be washed off with water and a little buffing. 2. The copper foil tape adhesive can overheat and boil out in spots where tape overlaps. Scrape it off with a craft knife and wash with water.
THERE'S A STICKY RESIDUE ALONG MY SOLDER LINE.	1. Foil not burnished well enough. 2. Foil got too hot.	1. Make sure you burnish the foil down very smoothly. 2. Reworking an area too much can overheat that section and cause the adhesive to boil out. 3. Remove adhesive residue with a craft knife and water.

PROBLEM	CAUSES	SOLUTIONS
THE SOLDERED FOIL HAS LIFTED OFF OF THE PIECE.	1. Overheating. 2. Overfluxing.	1. Heating an area up too much and re-working it can cause the adhesive to loosen and lift. When reworking an area, give it some time to cool. 2. Flux can seep under the edge, loosening the adhesive. Use your fingers or a pair of flat pliers to gently bend the metal back into place so that it lies flat.
THERE'S A LITTLE DIVOT IN THE SOLDER.	Adhesive oozing through.	A divot can occur when adhesive oozes up through the solder. Use a craft knife to clean it out, apply flux, and re-melt or add a little more solder.
THE SOLDER HAS A ROUGH TEXTURE WHERE IT HARDENED LAST.	Solder set too fast.	Hold the heat longer and give the solder more time to turn molten before removing the heat. Smooth out any rough areas with a file.
I CAN'T MAKE A MOUND OF SOLDER FOR THE JUMP RING.	Too much flux.	Don't add flux before making the mound. Remove excess flux with a cloth.
THE JUMP RING WON'T STICK IN THE SOLDER.	1. Not enough flux. 2. Wrong type of metal.	1. Put flux on the jump ring before melting solder to it. 2. Some metals won't accept melted solder. What these mystery metals are made of isn't usually specified on the product packaging.
A HOLE BURNED THROUGH THE SOLDER AND THE FOIL.	Iron held too long in one spot.	When reworking the solder, make sure it doesn't get too hot. Give it some time to cool. If a hole forms, peel off all of the foil and solder to start over. You could patch the hole with copper foil and solder, but the task isn't worth the trouble involved.
THE GLASS CRACKED WHILE I WAS SOLDERING.	1. Too much heat. 2. Uneven surface between glass. 3. Glass too thin.	1. When you cut the piece of glass, tiny cracks may have formed around the edge of it. Heat can cause these cracks to worsen. Sudden temperature changes—cold to hot or hot to cold—can also cause glass to crack. Make sure the edges of the glass are smooth and keep it at room temperature or warmer when soldering and cleaning it. 2. If the items under the glass aren't of a uniform thickness, the glass can rock and crack while you're handling or clamping it. Make sure the items are positioned in between the glass sheets as evenly as possible. Add spacers if necessary. 3. If the glass is too thin, it can easily crack. Use glass that's at least 1 mm thick.
THERE'S MOISTURE UNDER THE GLASS; THE INK RAN.	1. The seal isn't tight enough. 2. Wrong type of ink used.	1. Burnish the foil well and make sure it doesn't lift up after you solder. Try not to use too much flux or too much water when cleaning. The moisture will evaporate out over time. 2. Inkjet ink will run if it gets wet. Use a laser printer or photocopy machine for printed papers.
THE PAPER IS STUCK TO THE GLASS IN SPOTS.	Wrong type of paper used.	Use matte instead of glossy paper.

HARD SOLDERING

Hard soldering requires a butane torch and sterling silver solder. It's used to join sterling silver and other fine metals together. It can also be used to melt sterling silver solder and create decorative solder elements. Hard soldering projects don't involve glass and copper foil. The technique is used for creating metal projects, usually from sterling silver.

HARD SOLDERING WORKSPACE

Having a designated hard soldering area is ideal because you won't have to put your equipment and materials away each time you're finished with them. As with soft soldering, you should work in a standard-sized room that's well ventilated, where you have access to a sink and electrical outlets. Getting the room's lighting right is important. The flame of your torch will be easier to see in a room with dim lighting. I recommend using lamps instead of overhead lighting, or working in a room in which you can turn down the lights.

You'll want to work at a table that's sturdy and spacious. (Remember: accidents do happen even if you protect your work area, so using grandma's antique table in your hard soldering space is not recommended!) A chair whose height you can adjust will eliminate the need for hunching and add to your general comfort. Floor protection is a good idea, too—you don't want a hot tool to fall on your favorite rug!

You'll need separate work areas for various phases of the hard soldering process. A long table or a few tables set up near one another should suffice. You should have a designated area for doing metal work such as cutting, assembling, and preparing items to be soldered. You'll need a place for your pickle pot, which will have to be plugged in (see page 30 for more on pickling), and you should have a separate space for finishing work. Arrange your tools and materials in an orderly fashion in each of these areas. This will make the soldering process safer and more enjoyable.

In the area where you'll be soldering, place your tools on one side and your torch on the other, so that you won't be reaching across the hot flame for a pair of pliers. You should hold the torch in your non-dominant hand. This will leave your dominant hand free to do detail work with flux, tweezers, etc.

Work Surface

As with soft soldering, a large ceramic tile placed on a tabletop serves as my work surface. (Several tiles put together make for an even larger work area.) I keep the tile covered with a soldering board or a charcoal block— two of the many soldering surfaces that are available. The soldering surface you use will depend on the project at hand. Commercial soldering pads, ceramic pads, pottery bricks, paving bricks, pumice, and charcoal are all options.

Soldering boards are inexpensive and come in different sizes. Some materials reflect or absorb heat. Some are soft; others are firm. Hard and soft varieties are available. You can push a project into a soft board, which provides stability, and pin items in place. A hard solderite board is longer lasting but resistant to pins and pressure—you can't press items into a hard board. Both types reflect heat and are clean to work with. My soldering board is made of calcium silicate and has inert fillers and reinforcing agents that give it strength. It reflects heat well and cools down quickly.

Charcoal blocks are another inexpensive option. Soft and natural, they're made from wood that's been treated so that it will last longer.

Charcoal blocks absorb very little heat. Instead, they reflect heat, which makes the flame of the torch more effective. As with the soft soldering board, items can be pinned and pushed into a soft charcoal block for stabilization. The blocks do wear out faster than other surfaces, and it's a good idea to wrap a binding wire around the outside edge of a block to help hold its shape. Once a block starts to wear down, its surface can be sanded and made smooth again. Charcoal blocks can keep burning after soldering (you'll see a glow), and they can be a bit messy. Once a block starts to fall apart, you can re-use the scraps. Just place them in a container and save them for use with odd-shaped items that need to be stabilized while you're soldering them.

BASIC TOOLKIT FOR HARD SOLDERING

To make the hard soldering projects in this book, you'll need the following tools:

- » BASIC TOOLKIT FOR HARD SOLDERING
- » SOLDERING SURFACE (SOLDERING BOARD, CHARCOAL BLOCK, CERAMIC PAD, ETC.)
- » FLUX
- » SOLDER
- » SOLDER PICK
- » TORCH
- » PICKLE
- » SLOW COOKER
- » PLIERS
- » TWEEZERS
- » WIRE CUTTER OR JEWELER'S SAW
- » FILES
- » SANDPAPER OR ROTARY TOOL
- » BRASS-BRISTLE BRUSH

Torch

You'll need a standard hand-held micro or jumbo butane for hard soldering. The torches are available at most hardware stores. They're refillable (canisters of butane are also available at most hardware stores) and use no electricity. The micro torch works great for small jobs such as jump rings. The jumbo torch is good for larger projects and thicker metal. The torches cost from $30 to $70. Professional jeweler's torches start at around $500. Begin with a less-expensive model before investing in a pricier one. You can create all kinds of projects with standard butane torches, and you may never need to upgrade.

> SAFETY: *Light a torch only when you're using it. Never leave a lit torch resting in its stand. Point the flame toward the solder board, away from you, when lighting it. Give the torch a rest after about 30 minutes of continuous use.*

Hand Tools

PLIERS

As with soft soldering, you'll need various pliers, including concave bending pliers and flat needle-nose pliers. Round-nose pliers will come in handy also. They'll allow you to hold onto things while soldering and to shape the wires and the pieces for your projects. These tools will get messy, so keep your best beading pliers (or anything else you want to keep clean) away from the soldering area.

TWEEZERS

Use both locking and non-locking tweezers. Tweezers made of stainless steel are best for soldering. I prefer bent-tip, cross-locking tweezers with fiber grips that protect my fingers from heat. These are perfect for holding items while soldering. Since they lock, my hand doesn't fatigue as quickly when I'm holding them. Metal tweezers will heat up, so be careful when using a pair without grips. Metal tweezers with a fine tip are useful for picking up small pieces of solder and putting them in place. Copper tweezers or tongs are necessary for dipping soldered pieces into the pickle. You can never have too many kinds of tweezers lying around!

An "extra hand" tool has a heavy base and a holder with a pair of locking tweezers. The tweezers act as a third hand. The holder is jointed so that the tweezers can be easily moved into different positions.

SOLDER PICK

A solder pick is used to move and pick up hot solder or pieces of metal. Choose one that's made from titanium, because solder won't stick to it.

WIRE CUTTERS/JEWELER'S SAW

You'll need light- and heavy-duty wire cutters to cut pieces of metal and solder. You can also use a jeweler's saw or handsaw, but proper control of this tool takes a lot of practice.

Finishing Tools

Filing and sanding (finishing) will bring a nice shine to a piece. The easiest way to finish a project is with hand files or sandpaper. Start with a heavier grit and work up to the finest grit; then bring the solder to a shine with a brass-bristle brush. Hand soap and water will brighten the piece up even more.

| Hard soldering tools: (1) copper tweezers, (2) butane torch, (3) sandpaper, (4) nail file, (5) sanding and finishing bits, (6) metal sanding file, (7) metal diamond file, (8) rotary tool, (9) ceramic tile, (10) popsicle stick, (11) texture hammer, (12) solderite board, (13) charcoal block, (14) jewelers' extra hand with tweezers, (15) brass bristle brush, (16) needle-nose pliers, (17) large wire cutters, (18) titanium solder pick, (19) locking tweezers, (20) locking needle-nose hemostat pliers, (21) slow cooker. |

Rotary tools will make finishing work go much faster, and you can choose from a number of sanding and finishing bits. To create a smooth, buffed surface, I recommend using a finishing stone imbedded with fine emery abrasives. A sanding drum will take care of really rough spots.

HARD SOLDERING MATERIALS

Solder

Many different types of silver solder are available. The solder is made of a mixture of silver and zinc. The ratio of the proprietary mix allows different types of solder to melt at different melting points. Silver solder is sold in wire form, in sheets designed to be cut into small chips, and in paste form in a jar. The type of solder you use—wire, sheet, or paste—is up to you.

Solder is classified according to hardness and ranges from hard to medium to easy and extra easy. The hardest solder has the highest silver content and must be melted at a high temperature. It gives the best-looking silver finish.

The varying degrees of hardness prevent a project from falling apart when you have to solder it in multiple places. If you were to use the same hardness of solder throughout a project, you might melt the joints you made earlier as you progress. So it's smart to start the first joint with a hard solder and the next one with medium. You can then move to easy and even extra easy, if necessary. The diminution of hardness makes it possible for the flame to be held in place for a shorter amount of time in order for the solder to melt. If you have more than four solder joints, you may need to protect the previous solder joints with a heat sink, such as tweezers, or arrange your piece in a soft soldering block so the joints will not fall apart when re-melted.

Flux

Flux—a necessity for soft soldering—is also required for hard soldering. Flux allows the solder to flow by minimizing oxidation. It also holds small pieces of solder in place while you're working. Many different kinds of flux are available, including self-pickling and fire scale retardant types (see below for information on pickling). I recommend a basic paste flux made from Borax and water, which can be bought pre-made (you can also make it yourself, if you prefer). Keep a tight lid on the flux. If it dries out or becomes too thick, add some water to it. Apply flux with a small brush.

When flux is heated, it changes from a pasty white substance into a clear glaze that's very sticky; this sticky flux can hold a piece of solder in place on the surface of your piece. As the glazed flux cools, it hardens and can "glue" your project to your soldering board Glazed flux must be removed with pickle, whereas unglazed flux can be removed with water.

Pickle

Pickle is a mildly acidic solution that comes in granular form and is used to remove flux and surface oxidation. To make pickle, combine the granules with water and heat up the mixture in a small slow cooker. Turn your slow cooker on when you arrive at work, so the pickle will be nice and hot when you are ready to use it. Do not leave projects in the pickle for long periods of time, as the acid can weaken solder joints. You should only use copper

Hard soldering materials: (1) butane, (2) sterling silver bezel strip, (3) silver wire solder, (4) paste flux, (5) pickle, (6) silver sheet solder, (7) liver of sulfur, (8) sterling silver sheet, (9) sterling silver wire, (10) sterling silver jump rings, (11) jewelry findings.

tongs when working with pickle. Don't use steel tools, because the steel can contaminate the pickle and cause your project to become discolored.

> SAFETY: *If the pickling solution gets on your hands, you may experience a slight burning sensation similar to the way lemon juice feels in a cut. If this happens, just wash your hands with soap and water. You should place your slow cooker on a glass plate to protect your work table from spills and splashes.*

Extras

SILVER

Sterling silver is available as a raw material in sheet, wire, and strip form. Sterling silver is often referred to as 925 because it has 92.5% silver content and 7.5% copper. Other types of silver have different percentages of content. Fine silver, which is very soft, is 99.9% silver. It's not a good material to use alone for jewelry projects. Silver is usually sold by weight, and its cost depends upon current market prices. Sterling silver can be pricey—remember this if you make a soldering mistake. Save any project that goes bad, and keep your scraps so that you can get your money's worth out of the material.

OTHER METALS

Copper, brass, and other metals can be combined with silver to make interesting projects. Keep in mind that the metal you solder needs to melt at a higher temperature than the solder itself. For example, pewter melts at the same temperature as soft solder and would completely melt away under the heat of the torch's flame.

JEWELRY FINDINGS

All kinds of jewelry findings made from metal are available, including ring shanks, bezels, and clasp bails. Any pre-fabricated finding of this type can be used on your soldered project as long as it's made from sterling silver or another high-temperature metal.

You'll use this same basic technique for every hard soldering project in this book: Move the flame around the entire piece to warm it and then move the heat to the joint and melt the solder. Remove the heat, quench the piece in pickle, rinse it with water, and apply finishing techniques.

Once you become accustomed to working with the torch and develop a feel for the flow of the solder, you'll make fewer mistakes and find that the process gets easier. You'll also be able to figure out which type of soldering works best for different projects. Practicing soldering samples to get a feel for heating the metal and getting the solder to flow before you start your first important project.

Using the Torch

Start with a neutral flame: adjust the nozzle of the torch until its hole is half open and you see a bright blue cone in the middle of a translucent flame (see the photo below). The end of that cone should appear feathered. Just beyond this feathered end is the ideal spot to use for heating solder.

The addition of oxygen will cause the flame to get hotter; a very hot flame is called an oxidizing flame. The reduction of oxygen results

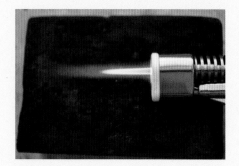

in a cooler flame that's known as a reducing flame. Oxidizing flames are used for fast heating, but—as the name implies—the flame can oxidize the metal. The reducing flame, which isn't hot enough to melt solder, is used for annealing or softening metal.

You'll be using a flame that falls in between the oxidizing and reducing states to melt and bond your solder. Controlling the flame distance and the movement of the heat is very important, as these factors ensure that the solder flows correctly and the heat doesn't melt or damage your piece. The colors of a heated piece can go from pink to pale red and red, and then to orange. You want to be in the pale-red or red for soldering. Once your metal turns red, it is about to melt, so you should remove the heat from your piece immediately. This is where dim lighting in your workspace may prove to be beneficial—you need to be able to determine the color of your heated piece.

Chip Soldering

Chip soldering uses pieces of solder that are cut from a sheet of the material. The sheet of solder should first be cut into thin long strips 1 to 2 mm wide. The strips should then be cut crosswise into little pieces. The pieces don't need to be the same size because you'll need very small ones (the size of tiny seeds) as well as slightly larger ones (the size of rice). Cutting the sheets ahead of time is a good idea, but make sure you put the pieces in containers that are labeled according to hardness.

1 Start with clean tools and materials. Remove any baked-on, sticky flux from your work surface and tools with a sanding file. Use rubbing alcohol or dish soap to remove tarnish, dirt, oils, or fingerprints from the metal you'll be soldering. You can also clean the metal by dipping it in the pickle. Use pliers or tweezers to handle the metal.

2 Position the piece to be soldered, making sure that it's stable and won't move around as you're working on it.

3 Apply flux to the metal. The area that's being soldered shouldn't have any gaps, and the joint should be as tight as possible. Make sure you coat all the metal on the project with flux where the flame will touch it to help prevent firescale.

NOTE: Firescale appears as a subtle reddish blotchy area on sterling silver. This happens when the copper content in the sterling silver rises to the surface during heating. Firescale is very difficult to prevent completely. Firescale is only removed by sanding the thin layer of copper from the surface of the metal.

4 Pass the flame back and forth in a circular motion over the fluxed project to warm it, until it looks like it has a clear glaze.

5 Place a small piece (roughly 1 to 2 mm) of precut solder onto the warm sticky flux using a pair of tweezers or your solder pick. The project should be warm but not too hot. You don't want it to be glowing red—this would cause the solder to flow too liberally and the metal to potentially begin melting.

6 Heat the entire piece again evenly, bringing the flame in closer as you move it around. Start at the back of the project away from the joint and then move the flame toward it. Heat the thicker spots longer than you do the thinner ones. Move the heat to the joint section where you placed the solder. The metal should be light pink in color and the flux should appear glossy. Immediately after the solder flows, remove the heat.

7 Move the piece with tweezers before it cools. Otherwise, the flux may harden and cause the project to stick to the soldering board.

If the flux does harden, don't pry the project off the board. Instead, lightly heat up the flux to loosen it and remove the project.

8 Submerge the project in the warm pickle. Be careful—the pickle may spray off the hot metal. You can cool down the metal before pickling it by immersing it in a bowl of water first. This is called "quenching." Leave the project in the hot pickle until the metal is clean of all oxidization and flux. Rinse your piece with water.

Pick Soldering

Pick soldering is a technique used for small areas of a project that need very little solder and in cases where using tweezers is too difficult. It produces a nice solid joint with very little—if any—solder to remove.

1 Start with a clean soldering pick. Remove any baked-on flux or solder with a file.

2 Place a cut piece of solder on the soldering board and coat it with flux.

3 Melt the solder chip until it's a shiny ball.

4 As you remove the flame, place the tip of the pick on the ball. It should stick to the end but not melt onto it. You should be able to knock it off with some pressure.

5 Flux and heat your metal project as described above until the flux is clear and sticky.

6 Keeping the metal warm, place the solder ball on the joint. The ball should slip off easily and stick to the flux on the joint. If the ball bounces off, the metal isn't hot enough.

7 Heat the piece and the solder until the solder flows. Remove the heat.

Sweat Soldering

Sweat soldering is used to join two flat pieces of metal together when you can't get the torch to the solder spot. With this technique, you'll heat the pieces so that the solder sweats from one piece of metal to the other and connects them.

1 Flux one piece of metal and heat it until the flux is clear and sticky.

2 Place the solder where the two pieces will be joined and then heat the piece until the solder begins to flow.

3 Coat the second piece with flux in the spot where the first piece will be joined to it.

4 Place the first piece with the melted solder on it on top of the second piece, where the pieces join.

5 Heat both pieces together until you can see or feel them sink or settle together. Use a solder pick if needed to hold the top piece in place.

Finishing Techniques

When you've finished soldering and have cleaned your project in the pickle, you can finish it up with sanding, texturing, polishing, and patina.

SANDING

Sand hard soldered projects the same way you do soft soldered ones—by hand or with a rotary tool.

POLISHING

For hand polishing, use a brass-bristled brush to bring the silver to a high shine. You should use a little soap and water with the brush to keep the bristles from leaving a yellowish tint on your silver. You can also shine your project with a soft cloth and some silver polish or buff it with a rotary tool. Don't press too hard when polishing, especially with the rotary tool or you'll remove details from the piece.

TEXTURING

The rotary tool makes texturing very easy, although it can be done by hand as well. To create brushed or textured effects on metal, you can use a sanding bit or wheel, a diamond bit, or sandpaper.

PATINA

To give your soldered piece an antique finish, apply a patina such as liver of sulfur. This patina makes different colors on the surface of silver. It's made of potassium sulfides and is most commonly available in lump form. The lumps have a distinct sulfur smell, similar to rotten eggs. Dissolve a lump of liver of sulfur in water. Submerge your soldered pieces in the solution, until you achieve the desired color. The colors that result range from pale rose to black, and their appearance depends upon the length of time the piece sits in the solution. When you're satisfied with the desired color, rinse the piece with cold water. You can sand the patina so that the darker color remains in the recessed areas of the design and gives the piece depth.

SAFETY: *Patinas are made with very strong chemicals, so be sure to wear gloves and safety glasses when working with them. Work in a well-ventilated area.*

PROBLEM	CAUSES	SOLUTIONS
THE SOLDER ISN'T FLOWING.	1. Not enough heat. 2. Too little or too much flux. 3. Dirty metal.	1. Make sure the flame is in a neutral position and that there's enough butane in the torch to keep it going. 2. Use a nice, even coating of flux. Add more flux or remove it by putting the project in the pickle again. 3. If the metal is too tarnished or dirty, it needs to be cleaned. Try buffing it.
THE SOLDER WON'T STICK TO THE SOLDER PICK.	Dirty solder pick.	Clean the solder pick by sanding it.
THE PIECE STUCK TO THE SOLDERING BOARD AFTER SOLDERING.	Flux has solidified, gluing the project in place.	Reheat slightly to loosen the flux. Remember to move your piece once the solder has set to prevent it from sticking.
THE SOLDER WON'T FLOW INTO THE JOINT.	1. Joint is not closed all the way. 2. Uneven heating pattern.	1. Make sure the joints are touching and closed all the way. Solder will not fill in the gaps. 2. Heat both the joint and the solder.
THE SOLDER FLOATS OUT OF PLACE.	1. Flux hasn't been heated. 2. Flux has cooled.	1. Heat the flux first until it's a clear glaze and becomes sticky. 2. Don't wait too long to place the solder, or the flux will harden. Reheat, if necessary.
I CAN SEE THE JOINT LINE AND THERE'S A DIVOT.	1. Not enough solder. 2. Not heated long enough.	1. Make sure you use the right amount of solder for the job. 2. Pull the flame away when you see the joint is smooth and filled in.
MY METAL MELTED BEFORE MY SOLDER DID.	Wrong metal or solder used.	Solder and metals can look the same. Make sure you use the right ones. Mark your solder and your metals with a permanent pen or keep them organized in marked containers.
MY FIRST JOINT CAME UNSOLDERED WHEN I DID MY SECOND JOINT.	Same solder hardness was used.	When doing several joints on one project, use harder solder and work down to easy. As you solder each successive joint, less heat will be required, and the risk of melting previous joints will be lessened.
MY SILVER TURNED GRAY AND BLOTCHY.	Firescale or fire stain.	Fire stain occurs during the heating process. Dipping the silver in the pickle is usually enough to remove it. Deeper gray or reddish stains may also be firescale and should be sanded off. Use a coating of flux to somewhat protect the silver, and don't overheat the metal.
MY SOLDERED PIECE IS IN THE WRONG PLACE.	The piece floated out of place or was moved before the solder set.	To unsolder, paint with flux, reheat the piece, and move the piece into the correct place once the solder flows. If necessary, remove the piece, clean it in pickle, and start the soldering process over.

BEADING BASICS

TECHNIQUES FOR WORKING WITH WIRE

If you're an accomplished jewelry maker, you already know the joy and creativity involved with putting together your designs. For those who are just starting, take heart: the process of learning is fun, and it's pretty simple once you master the basics.

Here are a few techniques to practice before diving into the jewelry-making projects later in this book. Refer to these steps later and review them as needed.

| Figure 1 |

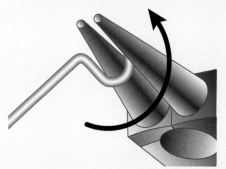

| Figure 2 |

| Figure 3 |

HOW TO MAKE SIMPLE LOOPS

Start with 6 inches (15 cm) of wire and work with a pair of round-nose pliers. Make a sharp 90° bend about ½ inch (1.3 cm) from one end of the wire, as shown in Figure 1. This measurement will vary, depending on how large a loop you want to make; with practice, you'll get to know how much wire to allow for it.

Hold the wire so that the longer portion points to the floor and the short, bent end is pointing at you. Grasp the short end with the round-nose pliers, holding the pliers so that the back of your hand faces you. The closer to the tips you work, the smaller the loops you can make. Keeping the tips themselves stationary, rotate the pliers up and away from you (see Figure 2). Be careful not to pull out the right-angle bend you made earlier. Stop rotating when you've made half the loop.

Slide the pliers' tips back along the wire a bit and resume the rotation. To prevent the loop from becoming misshapen, make sure to keep one of the pliers' tips snug inside the loop as you make it, so that the loop is being formed by a combination of rotation and shaping around the "mandrel" of the pliers. Keep working, sliding the pliers back as needed, until the loop is closed against the 90° bend (see Figure 3).

A bead loop is made by enclosing a bead between two loops. Another option is to start with an eye pin, so that you'll have to fashion only one closing loop. This method is used a lot in this book.

Of course, there are hundreds of variations of these basic links. Loops and links can be attached to each other with jump rings or linked directly together as you make them.

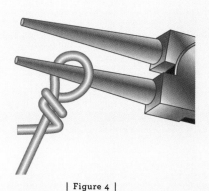

| Figure 4 |

A wrapped bead loop is a simple variation of the bead loop. Use an extra length of wire for the 90° bend. Once you've made the loop, reposition the pliers so that the lower jaw is inside it. Use your other hand to wrap the wire's tail around the base of the loop several times, as shown in Figure 4. Slide on one or more beads and, if the design calls for it, repeat the loop-forming process at the other end to make a wrapped bead loop (see Figure 5). Trim off any excess wire.

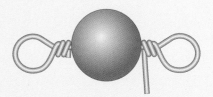

| Figure 5 |

OPENING AND CLOSING LOOPS

Just as with jump rings, use a pair of pliers to open and close loops. Twist the cut end sideways while keeping the other side of the loop stationary. As with jump rings, pulling it open any other way will distort the loop's shape. Be sure to tighten any gaps in loops after you've attached your links.

| Figure 6 |

TWISTING

Only square wire can be twisted. Round wire won't show the twisting properly.

To create twisted wire in no time at all, work with a pair of pin vises. Insert each end of a piece of wire into a pin vise, tighten the chucks, and twist them in opposite directions until you like the look you've achieved. If you have only one pin vise, secure the other end of the wire in a clamp or table vise (or in a pair of pliers if you have just a short quantity to twist). You can also use this tool to twist two lengths of the same wire together, creating a heavier look, or to twist together two different colors of wire.

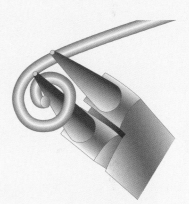

| Figure 7 |

MAKING SPIRALS

To create flat spirals, use the tip of a pair of round-nose pliers to curve one end of the wire into a half-circle or hook shape about ⅛ inch in diameter (see Figure 6). Use the very tips of the pliers to curve the end of the wire tightly into itself, aiming to keep the shape round rather than oval, as shown in Figure 7. Hold the spiral with flat-nose pliers and push the loose end of the wire against the already-coiled form (see Figure 8); as you continue, reposition the wire in the pliers as needed.

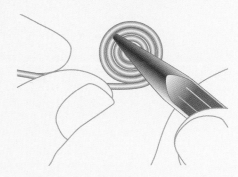

| Figure 8 |

PROJECTS

Soft Soldering

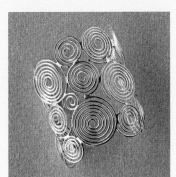

Hard Soldering

SOFT
SOLDERING

BEZELED MOONSTONE

DIFFICULTY

Easy

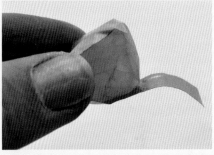

— A —

— B —

— C —

YOU WILL NEED

Soft soldering toolkit
Large moonstone
2 medium jump rings

Copper foil tape, $^5/_{16}$-inch (0.8 cm) wide
Lead-free solder

INSTRUCTIONS

1 Wrap copper foil tape all the way around the moonstone along the stone's smoothest, straightest edge, overlapping the ends of the tape a bit.

NOTE: The size of the tape you use will depend upon the size of the stone. *(photo a)*

2 Melt solder over the copper. *(photo b)*

3 Solder the jump rings to the top edge of the stone. *(photo c)*

4 String the bezeled stone onto the necklace or chain of your choice.

WIRE-ON-GLASS NECKLACE

DIFFICULTY

◆

Easy

YOU WILL NEED

Soft soldering toolkit

Glass circle, 1½ inches (3.8 cm) in diameter

2 glass circles, each 1 inch (2.5 cm) in diameter

2 glass circles, each ¾ inch (1.9 cm) in diameter

¼-inch (0.6 cm) silver-backed copper foil tape, at least 18 inches (45 cm) long

24-inch (61 cm) length of 18-gauge silver-plated wire

Concave round-nose pliers

17 jump rings, each 3 mm

15-inch (38.1 cm) length of chain

5-inch (12.7 cm) length of chain

TIP: *I trace the glass onto paper first and then draw the swirls on the tracings.. This helps keep the sizes and shapes uniform.*

INSTRUCTIONS

1 Wrap each piece of glass with copper foil tape. Solder all of the edges.

2 Cut all of the wire pieces to the length you need first so that your scrolls will be about the same size. Cut the large scrolls twice the length of the glass—e.g., 1-inch (2.5 cm) glass should have a 2-inch (5.1 cm) wire. Cut the smaller scrolls half the size of the glass. Bend accordingly using the concave round-nose pliers and trim excess wire off.

3 On the edges of each circle, tack-solder the scrolls in place over the top of the glass.

4 Create three larger scrolls for the 1½-inch (3.8 cm) piece of glass. Tack-solder them in place.

5 Solder the jump rings to the top and bottom of each circle.

6 Attach the 15-inch (38.1 cm) length of chain to the jump ring at the top of the large circle. Attach the 5-inch (12.7 cm) length to the jump rings at the bottom.

7 Attach one 1-inch (2.5 cm) circle to the other end of the 5-inch (12.7 cm) length of chain via the jump rings. Connect the two ¾-inch (1.9 cm) circles using jump rings also.

8 Connect the other end of the 15-inch (38.1 cm) chain to the top of the ¾-inch (1.9 cm) circle via the jump rings.

BIRD'S NEST FRAME NECKLACE

DIFFICULTY

Easy

YOU WILL NEED

Soft soldering toolkit
Copper foil sheet, 1 × 2 inches
(2.5 × 5.1 cm)
Brass wing charm
Round brass frame, 1½ inch (3.8 cm)
diameter

36-inch (91.4 cm) length of 21-gauge
silver-plated wire
3 white pearl beads, each 5 mm
2 jump rings, each 5 mm
18-inch (45.7 cm) strand of pearl beads

TIP: *When I work late at night, I often get ideas for projects but don't have all the elements on hand that I need. On one such occasion, with no stores open, I decided to make my own bird charm. Instructions for making the charm are below, but you can buy a pre-made charm at a craft store, if you prefer. Stamped brass frames like the one in this project can be found in the scrapbook section of most craft stores.*

INSTRUCTIONS

1 Draw a bird shape onto the paper-backed side of the copper foil and cut it out. Make it a size so the wing shape will go with it.

2 Peel the backing off and apply flux and solder to the back of the copper foil. The adhesive will melt away, although it may take several passes with the iron for it to do so. Scrape off any black chunks of adhesive that are left behind.

3 When the solder sets on the back of the copper foil, melt solder over the front side.

4 Use a clamp to secure the brass wing charm in place on the foil bird shape. Then melt solder over it to tack it in place. Cover the front side of the wing with solder so that it's the same color and texture as the bird's body.

5 Use a clamp to secure the bird to the frame. Then tack-solder it to the frame from the back side.

6 To make the bird's nest, string three pearls onto the end of the silver-plated wire, leaving a tail at the end that's about 18 inches (45.7 cm) long. Bend the beaded wire into a triangle and coil the remaining wire around it to form the nest. When you're done wrapping, pull the tail in through the center.

7 After you finish coiling the wire, twist the end of the coil and the tail together. Tuck this twisted end up under the nest.

8 Clamp the nest onto the edge of the frame and tack-solder it in place.

9 Solder the jump ring onto the top edge of the frame from the back side. Use another jump ring to attach the pendant to the pearl strand.

NOTE: To get a rustic look to the piece, do not wash off all the flux yet. Allow the flux to stay on part of the brass frame to cause it to oxidize, creating a verdigris look. This process can take a few days to achieve the desired look. Once the desired look is created, rinse the remaining flux from the piece. Do not sand or polish, or you will remove the verdigris.

OWL PENDANT

DIFFICULTY

✦

Medium

YOU WILL NEED

Soft soldering toolkit
Crown-edge bezel cup, 32 mm
Head pin
2 daisy spacer beads, each 2 mm
48 inches (1.2 m) of 20-gauge
   silver-plated wire

Owl charm
Large round metal bead
Small round metal bead
36-inch (91 cm) strand of leather

NOTE: A bezel cup is used in jewelry making for setting stones.

TIP: *If you don't have a bezel cup, use wire or a different metal jewelry finding for the bottom of the cage.*

INSTRUCTIONS

1 If the bezel cup has a hole in the middle, place the head pin with a daisy spacer on it through the hole. To secure the pin, create a loop on the underside of it with needle-nose pliers. You can hang a bead from the loop, if you like.

2 Cut the silver wire into eleven 4-inch (10.2 cm) pieces.

3 Using round-nose pliers, curl both ends of each wire piece into small, closed loops.

4 Bend each piece of wire into a U shape with the pliers.

5 Solder the six wire pieces onto the edge of the bezel cup so that the ends of each piece are directly across from each other. Space them evenly around the edge of the bezel cup and arrange them so that they all cross one another at the very top

in the middle. Crimp the center of the wires at the top as needed to accommodate the addition of new layers of wire.

6 Create a wire hanger for the owl by making a loop that attaches to the charm, twisting the wire with pliers to close the loop and then making a second loop with the free ends. One end should terminate with that second loop, while the other end should extend another 1½ inches (3.8 cm) or so beyond it. Hang the owl from the top center of the crossed wires to secure it.

7 String the beads and the second spacer onto the long end of the wire, bending the wire as needed to make it point straight up from the middle of the cage. Be sure to leave enough space between the bead and the cage so you can solder the wire onto the top of the cage. At the top, above the beads, create a

wrapped loop. Solder the wire below the beads onto the cage.

8 Form a small ring about ⅝ inch (15 mm) in diameter from a scrap of wire. Solder the ends together to make the door to the cage.

9 Use a locking hemostat plier or alligator clamps to hold the wire in place. Arrange the ring so that it touches three cage wires— one in the middle, and two on either side. The outer cage wires should just touch the sides of the ring. Solder the ring onto the cage wires.

10 Once the ring is in place, cut the center wire out of the circle to create the door.

11 Hang the pendant on the leather strand.

BEADED BEZELED STONES

DIFFICULTY
✦
Easy

YOU WILL NEED

Soft soldering toolkit
3 square stone beads, each 1¼-inch
(3.2 cm)
18 jump rings, each 2 mm
Flexible sterling-silver beading wire,
.36 mm in diameter

68 round beads, each 4 mm
32 round beads, each 6 mm
15 round beads, each 9 mm
24 silver crimp tubes
Lobster clasp

INSTRUCTIONS

1 Wrap each stone bead with copper foil tape, making sure the foil folds down over the front and back sides to create a secure bezel.

2 Solder over all of the foil tape.

3 On two of the stone beads, solder three jump rings on each of two opposite sides. Space the jump rings evenly, one in the middle and one on each end of the stone bead's side.
(see photos)

4 On the third stone bead, solder three jump rings on each of two adjacent sides.

5 Starting with the third stone, string beads onto each jump ring in a symmetrical pattern. Start each strand of beads by looping a 4-inch (10.2 cm) length of the beading wire onto a jump ring and securing it with a crimp tube. Trim the end of the wire as needed. The innermost strand of beads should be the shortest of the three (approximately 2.5 inches, or 6.3 cm), which you string to the top of the stone. This will ensure that the two outer strands—approximately 3 inches (7.6 cm) for the middle strand and 3.5 inches (8.9 cm) for the longest outer strand—drape in a curve.

6 Cut a 6-inch (15.2 cm) length of the beading wire. Attach it to the bottom jump ring on one side with a loop and a crimp tube. Then string the beads symmetrically and secure the beading wire to the bottom jump ring on the other side in the same way.

7 Finish the end of each bead strand with a crimp tube to secure it to a corresponding jump ring on one of the first two stones.

8 For the top strands, each strand attaches to the top of the first two stones. Cut the beading wire to a length of 8 inches (20.3 cm). Repeat the process for the first six strands, making the next six identically symmetrical.

9 Gather the three long ends of the 8-inch (20.3 cm) strands on one side of the necklace and string them through a crimp tube. Adjust the strands so they hang nicely, and crimp the tube tightly in place. Repeat the same step on the other side.

10 String the rest of the beads onto the remaining sterling silver wire. Secure the ends with crimp tubes and attach the lobster clasp closure.

TIP: *Stone beads will become much more versatile thanks to your new soldering skills! The beads often come with only one hole drilled into them. By adding your own bezel and jump rings, you'll have limitless beading-pattern options.*

BRASS & COPPER
MATCHING SET

DIFFICULTY
✦
Easy

Soft soldering toolkit
2 pieces of 24-gauge copper-finish
    textured sheet metal, each
    2 × 4 inches (5.1 × 10.2 cm)
2 pieces of 24-gauge brass-finish
    textured sheet metal, 2 × 4 inches
    (5.1 × 10.2 cm)

2 premade metal wire bangle bracelets
Nylon bracelet-bending pliers
16-inch (40.6 cm) length of
    18-gauge wire

TIP: *Premade wire bangle bracelets are very inexpensive, and they're perfect for jewelry-making projects. The textured metal sheets have finishes on them that make them difficult to solder. You should sand the areas of the sheets that you plan to solder to remove the finish and make the metal easier to work with.*

WRIST CUFF INSTRUCTIONS

1 Using heavy cutters, cut the four metal sheets into 1 × 2-inch (2.5 × 5.1 cm) strips. Set three pieces aside.

2 Sand the edges and round the corners of each sheet using a rotary tool with a sanding bit.

3 Trim 2 inches (5.1 cm) off each end of the two wire bangle braceletss. Bend each cut bangle bracelet to create a cuff that fits your wrist.

4 Using the bracelet-bending pliers, curve four pieces of the cut metal sheets so that they fit flush on the curved bangle bracelets.

5 Solder the end of one of the bangle bracelets onto the back of one of the textured metal pieces, about ½ inch (1.3 cm) from the top.

6 Overlap the next piece, and solder it in place. Continue this process until all four pieces are in place.

7 Solder the other bangle bracelet onto the inside of the cuff about ½ inch (1.3 cm) from the bottom of the textured pieces.

NECKLACE INSTRUCTIONS

1 Use the bracelet-bending pliers to curve three of the remaining metal pieces lengthwise.

2 Shape the 16-inch (40.6 cm) wire into a circle with round-nose pliers. Then create a wrapped loop on one end of the wire and basic a hook on the other with the pliers.

3 Lay the wire on the back side of the textured metal pieces, about ¼ inch (6 mm) from one end of each piece, and solder it in place.

FLOWER COCKTAIL RING
OR PENDANT

DIFFICULTY

✦

Medium

YOU WILL NEED

Soft soldering toolkit
Paper and pencil
1 piece of 24-gauge silver-plated
   3 x 5-inch (7.5 x 12.7 cm)
   copper sheet
Metal hand shears

Rotary tool or heavy metal file
Nylon-jaw pliers
Ring shank
1-inch (2.5 cm) length of 18-gauge wire
3 ball head pins
3 clear glass beads

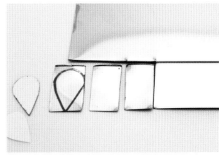

— A —

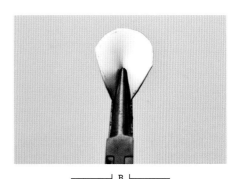

— B —

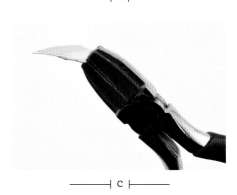

— C —

INSTRUCTIONS

1 Sketch a petal shape on a piece of paper to get a feel for the design. You'll need to decide on a size for the petals.

2 Determine the size of the metal pieces you need for each petal. The bottom ones should be a bit bigger than the top ones. Cutting the petals out of small pieces of metal is easier than cutting them out of the large sheet. I suggest that you cut small squares from the copper sheet, trace the petal shape onto them, and then cut out the petals using heavy metal shears. *(photo a)*

3 Using the rotary tool or metal file, sand the edges of each petal to remove any sharp or rough edges.

4 Use round-nose pliers to create a center crimp in each petal. *(photo b)*

5 Give the petals shape by using nylon-jaw pliers to bend their edges downward. *(photo c)*

6 Cut a circle that's ½–¾ inch (1.3–1.9 cm) in diameter from the copper sheet. The circle will serve as the base of the flower. Sand the edges of the circle to smooth them out.

7 Solder the largest petals to the base first. As the metal heats up, the plating on the copper will slowly burn away, leaving an interesting pattern.

8 Solder the top layer of petals onto the first layer, spacing the petals out to fill in the design.

9 Solder the ring shank onto the base of the flower.

10 If you want the ring to double as a pendant, bend the 1-inch (2.5 cm) wire into a loop with round-nose pliers and solder it onto the upper side of the base. *(photo d)*

11 Place each of the beads onto a head pin. Bend the end of each pin with round-nose pliers to secure the beads. Trim off the end of the pin, leaving enough metal so that you can solder it onto the flower. Solder each bead to the center of the flower. *(photo e)*

12 Gently bend each petal up with nylon-jaw pliers to give it a dimensional look.

| D |

| E |

TIP: *Most metals are composed of a brass or copper base with rhodium or silver-colored plating. This finish burns off when the metal is heated. If you want your piece to have a silver finish, use a sterling silver sheet or manually plate each piece of metal with solder before you solder the project together. You should plate the metal before you solder rather than after, because plating can make the joints melt and come apart.*

DECORATIVE BOTTLE TOPPERS

PAINTING THE BOTTLES

YOU WILL NEED

Assorted bottles
Glass paint

TIP: *Interesting bottles can be found at thrift stores and online. Small liquor and perfume bottles work great for this project, too.*

INSTRUCTIONS

1   Apply glass paint to each bottle according to the manufacturer's directions. Be creative! The paint I used was very thick. I wanted my bottles to have a sheer look, so I applied the paint and then wiped it off with a paper towel to achieve the transparency I was going for.

2   Let the paint dry. It can cure for 21 days, or you can heat-set it in the oven at 350°F (77°C) for 30 minutes and then let it cure for three days. Make sure you're satisfied with the painting and that the bottles are dry before you proceed with the project.

PURPLE BOTTLE

DIFFICULTY
◆
Easy

YOU WILL NEED

Soft soldering toolkit
Painted bottle
Large teardrop-shaped chandelier crystal
Lead-free came, ⅛ × 4 inches (0.3 × 10.2 cm),
   ³⁄₃₂-inch (2.4 mm) channel, flat U
4-inch (10.2 cm) strand of glass rhinestone chain
Black patina

INSTRUCTIONS

1 Wrap the top edge of the bottle opening with copper foil tape, covering it completely.

2 Solder over all of the copper foil.

3 Wrap the came around the chandelier crystal, trimming it so that the ends meet up at the point. Spot-solder the ends together, making sure you don't touch the tip of the iron to the came itself, as this will cause it to melt.

4 Solder the crystal to the center of the bottle top, melting a large mound first and sinking the crystal into it.

5 Cut the rhinestone chain so that it fits around the top of the bottle. Solder one end of it in place on the back of the bottle. Wrap the chain around the bottle snugly and solder the other end in place.

6 Apply black patina to all of the metal parts.

TIP: *Be sure to use rhinestone chain with glass rhinestones or crystals. Plastic rhinestones will melt when you're soldering.*

GREEN BOTTLE

DIFFICULTY
◆
Medium

YOU WILL NEED

Soft soldering toolkit
Painted bottle
7-inch (17.8 cm) length of silvered copper foil tape,
   ½ inch (2.5 cm) wide
Decorative paper
Aluminum foil tape
Flat-backed clear glass marble, 2½-inch (6.4 cm) diameter
Silver ball-end head pin
Silver daisy bead, 4 mm
Faceted glass bead, 10 mm

INSTRUCTIONS

1 Wrap the silvered copper foil tape around the neck and over the opening of the bottle. Cover the tape with solder.

2 Cut a piece of the decorative paper that will fit behind the flat-backed marble.

3 Cover the back of the paper with aluminum foil tape.

4 Place the paper image-side down on the back of the marble so that the image shows through its rounded front.

5 Wrap the edges of the marble and the paper with the silvered copper foil tape.

6 Solder over all of the foil-taped edges but not the back of the marble.

7 Working around the edge, melt solder balls around the outside of the marble.

8 Tack-solder the finished marble to the top of the neck of the bottle. You should tack-solder the marble from the back.

BLUE BOTTLE

DIFFICULTY

Easy

TIP: Flat-backed marbles can be found in all shapes and colors, and they offer a fun alternative to plain glass. The image on your decorative paper will be magnified under the marble.

Aluminum foil tape is great to use as a lightweight backing. Solder won't stick to it.

NOTE: Soldering the finished marble to the neck of the bottle will seal the bottle. None of the bottles in this project are intended to be opened; they are decorative.

9 String the head pin through the daisy bead and the faceted bead. Coil the end of the pin in place at the back of the faceted bead to create enough surface area so that you can solder it onto the neck of the bottle. Cut off the excess.

10 Tack-solder the coiled pin onto the front of the bottle by holding it in place and sticking the iron tip in from the side. Melt the solder that's already there so the coiled pin can stick to it. The bead is glass, so it won't melt.

YOU WILL NEED

Soft soldering toolkit
Painted bottle
Large crown charm
Copper patina

INSTRUCTIONS

1 Wrap copper foil tape around the neck of the bottle.

2 Drop solder balls around the neck.

3 Tack-solder the crown charm to the top of the bottle from the back.

4 Apply copper patina to the soldered edges and the charm.

TIP: Copper patina gives projects an antique look, which I think is lovely. The patina doesn't take well to lead-free solder, though. You have to reapply it until you get the color you want. When I made this project, the copper patina didn't take to the charm at all, so I covered it with a thin layer of solder and then reapplied it.

LAYERED FLOWER BROOCH/NECKLACE

DIFFICULTY
◆
Medium

YOU WILL NEED

Soft soldering toolkit
Assorted fabric swatches
12 pieces of cathedral window-shaped
    glass, each 1¼ × 1 inches
    (3.2 × 2.5 cm)
Silvered copper foil tape

3 head pins
3 faceted glass beads
Bar pin
Premade necklace
18-inch (45.7 cm) swatch of tulle

TIP: *You should be able to get fabric swatches for free at your local fabric store. Try to avoid material that's made of nylon—it can melt under the glass while you're soldering.*

INSTRUCTIONS

1 Cut six pieces from the fabric swatches. Sandwich each piece of fabric between two sheets glass to make six glass pieces. Wrap the glass sandwiches with copper foil tape and solder them around the edges.

2 Arrange two sets of three soldered shapes together so each set forms a triangle in the center. Solder each set of three together wherever the pieces touch.

3 Layer the two circles together and solder them wherever they touch.

4 Place a faceted bead on each of the head pins. To secure the beads, bend the wire at the end of each pin using round-nose pliers. Solder the bent pins onto the middle of the flower.

5 Solder the bar pin to the back of the flower near the end that will serve as the top.

6 Tie the tulle into a bow on the pre-made necklace.

7 Pin the flower to the tulle.

CHRISTMAS ORNAMENT

DIFFICULTY

✦

Easy

Soft soldering toolkit
Assorted decorative papers
2 pieces of glass, each 2 × 3 inches
   (5.1 × 7.6 cm)
2 pieces of oval glass, each
   1 × 1¼ inches (2.5 x 3.2 cm)
Wavy-edged copper foil tape, 5/16 inch
   (8 mm) wide
2 small solder scrolls
2 medium solder scrolls

Small heart solder scroll
3 small jump rings
Medium jump ring
5 head pins
Assorted beads
One-word charm
Large jump ring
6-inch (15.2 cm) length of beading wire
2 crimp tubes with jump rings attached

INSTRUCTIONS

1 Cut two 2 × 3-inch (5.1 × 7.6 cm) pieces of paper and two 1¼ × 1-inch (3.2 × 2.5 cm) oval pieces of paper to fit in between the sheets of glass. Each piece of paper should have a design that you want to feature in the ornament. Place the pieces of paper back to back between the sheets of glass.

2 Wrap the glass pieces with copper foil tape (I used wavy-edged tape for the bigger sandwiched glass piece) and solder the edges.

3 Solder the two small solder scrolls on the top edge of the 2 × 3-inch (5.1 × 7.6 cm) piece of glass, near the corners.

4 Solder the medium scrolls near the center of the bottom of the 2 × 3-inch (5.1 × 7.6 cm) piece of glass.

5 Tack-solder a small heart scroll where the two medium scrolls meet.

6 Solder three small jump rings to the bottom of the oval glass.

7   Loop the medium jump ring through the heart scroll and solder it to the top of the oval glass piece.

8   Make three beaded head-pin charms. Do this by stringing various beads onto a head pin and then creating a loop on the other end to hang them from. Attach the charms to the bottom jump rings on the oval piece of glass.

9   Create two more small head-pin bead charms for the top of the ornament and attach them to one of the solder scrolls at the top of the 2 × 3-inch (5.1 × 7.6 cm) piece of glass. Secure the bead charms by making simple loops at the end of each pin.

10   Attach the word charm to the heart scroll using the large jump ring.

11   String beads onto the beading wire to make a hanging loop. Secure the ends of the wire with the crimp tubes that have jump rings.

12   Attach the beaded wire to the solder scrolls on the top of the rectangular glass.

TIP: *Pre-made solder scrolls make projects go quickly while providing uniformity of size and shape. Be careful when you solder them in place, though—too much solder can flood the scroll and hide its details.*

HALLOWEEN CANVAS

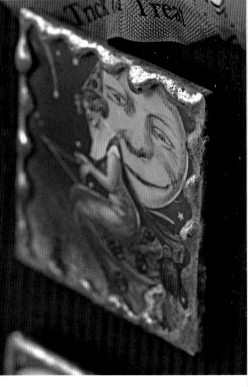

INSTRUCTIONS

1 Paint the canvas with the orange paint. Dry-brush a small amount of dark-brown paint onto the canvas to create a rustic look. Let the paint dry.

2 Cut a 7 × 7-inch (17.8 × 17.8 cm) square of plain decorative paper. Use the craft glue to adhere the paper to the center of the canvas.

3 Glue the decorative corners onto the corners of the paper.

4 Cut four images that you like from the decorative papers. Sandwich each one between two sheets of glass to make a total of four glass pieces. Wrap each one with copper foil tape. I used straight-edged tape for two of the pieces and wavy-edged tape for the others. Burnish tape.

5 Solder all of the edges.

6 Clean and polish each glass piece.

7 Use the heavy mounting tape to attach the four soldered glass frames to the paper in the center of the canvas.

8 Glue a bit of decorative ribbon to the top center of the paper.

TIP: *You can substitute papers with photos for a personalized canvas. If you do use photographs, print them onto matte-finish paper with a photocopier or laser printer.*

LEATHER CUFF

DIFFICULTY
✦
Medium

| A |

YOU WILL NEED

Soft soldering toolkit
Piece of leather, 1½ × 6½ inches
  (3.8 × 16.5 cm)
Decorative rubber stamp
Water-soluble ink
Bracelet-bending pliers
Copper foil tape, ½-inch (1.3 cm) wide

TIP: *Leather can make the tip of your iron dirty. Use an old tip for this project so that you don't ruin a good one.*

INSTRUCTIONS

1 Cut the leather to a length that fits your wrist.

2 Stamp a design onto the leather with the rubber stamp and water-soluble ink. *(photo a)* You can also trace a design or draw one freehand. To trace a design, print it onto regular paper and place the paper on top of the leather, print side up. Use a pen to trace the design, pressing hard so that an indention shows on the leather.

TIP: *Use a water-soluble ink, so if you get ink in a place that you don't want it, you can remove it easily. The ink won't run; it will get burned off easily.*

3 Run the tip of your hot soldering iron over the design to emboss it into the leather. *(photo b)*

4 Wrap ½-inch (1.3 cm) wide copper foil tape around the edges of the leather. The tape should overlap more on the back than on the design side.

| B |

5 Use the bracelet-bending pliers to bend the foil tape and the leather into a cuff shape. You can also hand-form the piece. Make sure you form it before you apply the solder. If you do it afterward, the solder and tape could crack or break apart. *(photo c)*

6 Solder over the copper foil tape.

| C |

PEACOCK FEATHER EARRINGS

TIP: *When you tack-solder, use very little solder so that it doesn't overflow onto the wire. You want to create the illusion that the glass is just sitting there.*

YOU WILL NEED

Soft soldering toolkit
2 peacock feathers
Aluminum foil tape 3 inches (7.6 cm) wide
4 glass circles, each 1½ inches (3.2 mm) in diameter

Copper foil tape, ½-inch (1.3 cm) wde
9-inch (22.9 cm) length of twisted, silver-plated 20-gauge wire
2 sterling silver ear wires

INSTRUCTIONS

1 Cut a large piece of the aluminum foil tape. Place one of the feathers front side up on the adhesive side of the tape. Repeat with the other feather. The tape backing will hold the feathers in place while you trim them. It will also help the designs of the feathers to show up and give the earrings a nice silver appearance on the back.

2 Place one of the glass circles over one of the feathers. Using the circle as a cutting guide, trim the feather so that it fits under the glass. Repeat with the other feather.

3 After trimming the two feathers, place two pieces of glass over each tape-backed feather. Wrap the edge of each glass-covered feather with copper foil tape. Solder over the foiled edges.

4 Cut the 9-inch (22.9 cm) length of wire in half.

5 Shape one of the pieces of wire to fit around one of the bottom of one of the soldered glass pieces with the feather inside. Bend the wire so that the ends cross one another at the top *(see photo)*. Do the same with the other piece of wire for the second piece of glass.

6 Form a loop with the ends of one of the wires and solder it closed. Repeat with the other piece of wire.

7 Place the soldered glass pieces in the bottom of the wire loops and tack-solder them in place at the sides and bottoms.

8 Attach the ear wires.

TIP: *Chunky charms can be tricky to work with. When making this project, try using charms that are of an even thickness. Otherwise, make sure you add spacers to the piece so that the glass sits flat. Spacers can be made from just about anything as long as they help the glass to sit evenly. Thicker glass works best with thicker charms.*

CHUNKY KEYCHAIN

DIFFICULTY
✦
Medium

Soft soldering toolkit
Assorted decorative papers and crystal
  stickers
2 pieces of glass, each 1¼ inches
  (3.2 cm) square and 2 mm thick
2 pieces of glass, each 1½ inches
  (3.8 cm) square and 2 mm thick
2 pieces of glass, each 1 x 1½ inches
  (2.5 x 3.8 cm) and 1 mm thick
2 large charms
Craft glue

Piece of glass, 1 × 1½-inch
  (2.5 x 3.8 cm)
Silvered copper foil tape, ½ inch
  (1.3 cm) wide
4 medium jump rings
Small jump ring
Lock-and-key charms
Rhinestone in bezel
3-inch (7.6 cm) length of wide-link chain
3 large jump rings
Key ring

INSTRUCTIONS

1 Cut your paper so that it fits
   the sheets of the glass. You'll
   need one piece of paper that's
   1¼ inches (3.2 cm) square, one
   that's 1½ inches (3.8 cm) square,
   and one that's 1 × 1½ inches
   (2.5 x 3.8 cm).

2 Decorate the paper pieces with
   crystal stickers.

3 Glue the large charms in place
   on the paper pieces.

4 Place the embellished papers
   between the pieces of glass.
   If your charms have uneven
   surfaces, use spacers so that
   the glass that covers them sits
   evenly and doesn't rock. For the
   crown charm, I used thick crystal
   stickers as spacers, placing them
   in each corner so that they add
   to the design. Spacers can also
   be hidden in the corners with
   tape and solder.

5 Wrap the edges of the
   sandwiched glass pieces with
   the silvered copper foil tape. The
   tape will be visible in the gaps
   between the pieces of glass
   inside the soldered squares.

6 Solder a medium jump ring to
   the top of each glass assembly.

7 Make decorative solder balls
   in each corner of the crown
   piece. (See page 22 for more
   on solder balls.)

8 Solder a small jump ring to the
   bottom of the rectangle piece.
   Attach the lock-and-key charms
   there using a medium jump ring.
   Solder the rhinestone bezel to
   the top of the rectangle. (If the
   rhinestone is made of plastic,
   carefully lift the prongs of the
   bezel and remove it. You can
   then solder the bezel. Once
   everything has cooled, place
   the stone back in place.)

9 Clean and polish each glass
   piece. Then attach them to the
   wide-link chain with the large
   jump rings.

10 Attach the chain to the key
   ring. If you want the project to
   have extra dazzle, add another
   keychain, as I did.

SHELL-ENCRUSTED TIN

DIFFICULTY

◆

Easy

YOU WILL NEED

Soft soldering toolkit
Tin box
Assorted shells
Copper foil tape, ½-inch (1.3 cm) wide

Assorted sea glass
Black patina

| A |

| B |

INSTRUCTIONS

1 Arrange the shells and the glass to see how they will fit together and to note the contact points of the shells where you need to apply foil tape.

2 Wrap the sides and bottoms of each shell and piece of glass with copper foil tape, keeping in mind that the tape needs to contour in slightly so that it will hold the objects in place once they're soldered. Solder over all of the copper foil. *(photos a–b)*

3 Apply a layer of solder over the top of the tin. *(photo c)* If the tin is resistant to the solder, you may need to sand some of the finish off. Keep heating and fluxing until a nice layer of solder sticks to the lid.

4 First solder each wrapped shell and piece of sea glass individually. Next place the soldered shells and sea glass onto the soldered lid and metal-solder around the shells to tack-solder them in place. *(photo d)*

5 Rinse the whole thing with water to remove all of the flux and coat the metal with black patina. You can apply patina to the entire project, but keep in mind that it may take to different types of metal differently. On this project, for example, it didn't show up on the box itself as well as it did on the solder.

TIP: *You can purchase seashells at a store or collect them in advance during a beach trip. If you get the shells during a trip, the tin can serve as a special souvenir or vacation keepsake.*

| C |

| D |

RING ASSORTMENT

FLOWER RING

FILLIGREE RING

DICE RING

FACETED STONE RING

MOTHER-OF-PEARL RING

BRACELET LINK

Soft soldering toolkit
Pendant or charm
Filigree charm
Ring shanks
Die
Copper foil tape, ½-inch (1.3 cm) wide
and ¼-inch (6 mm) wide

Faceted stone
Mother-of-pearl button
24-gauge silver-plated bead wire,
3 inches (7.6 cm) long
Round flat pearl with a bead hole
Decorative link from a bracelet

TIP: *Just about any object can be used to make an interesting ring. Keep your eyes open for unique found objects. Rings are fun and easy to create.*

INSTRUCTIONS

Crown, Flower, and Filigree Rings

1 Place the pendant or charm facedown on your soldering surface and solder the ring shank to the back of it.

2 If the pendant or charm was made to hang on a necklace or bracelet, build the ring so that the hanging loop is at the top. That way, the item can double as a pendant or charm and the ring shank won't be in the way. *(photo a)*

Dice Ring

1 Wrap ½-inch (1.3 cm) wide copper foil tape around the middle of the die and cover it with solder. Be careful—dice are made of acrylic and can melt under the heat from the soldering iron.

2 Using the needle-nose pliers, open the ring shank and bend the ends so that they form a U shape.

3 Solder each end of the U shape to the sides of the die where the solder strip is. *(photo b)*

Faceted Stone Ring

1 Wrap copper foil tape around the edges of the stone, folding it over so that it covers the back completely.

2 Solder the back first, covering all of the copper foil. Turn the stone over and solder the sides.

3 Secure the stone facedown with a vise so that it doesn't rock while you solder the ring shank to the back of it.

⊢ A ⊦

Mother-of-Pearl Ring

1 Wrap ¼-inch (6 mm) wide copper foil tape around the outside edge of the button and cover the foil with solder. This step doesn't add to the structural integrity of the ring, but it does give it a nice look.

2 Run a thin piece of wire through the hole in the pearl and then through the buttonholes. Twist the ends together on the other side with the round-nose pliers. Bend the wire up so that it lies flat against the back of the button.

3 Melt solder over the wire.

4 Place the button facedown on your soldering surface and solder the ring shank to the soldered wire on the back.

Bracelet Link

1 Solder the link shank to the back side of the link, centering the link on the shank.

2 Place the link facedown on your soldering surface and solder the ring shank to the two soldered points you just created.

———| B |———

PRINCESS TIARA

DIFFICULTY

Easy

YOU WILL NEED

Soft soldering toolkit

Metal hair comb

13 inches (33 cm) of heavy, half-round wire

2 glass crystals, each 1½ inches (3.8 cm) square

2 glass crystals, each 1¼ inches (3.2 cm) square

Three glass teardrop chandelier crystals, each 1½ (3.8 cm) long

34-inch (86 cm) length of U-flat lead-free came, ³/₃₂ x ⅛-inch (2.3 x 3 mm)

6 large faceted glass crystal beads

3 decorative head pins

10½-inch (26.7 cm) length of 18-gauge wire

3 small silver round beads

3 flat headpins

TIPS: *This project can easily be expanded with more crystals and embellishments to make full crowns!*

Instead of using crystals, make glass collages and use other found objects to create themed tiaras and crowns.

Combs aren't necessary. You can bend the wire into a curve instead and secure the tiara with bobby pins.

INSTRUCTIONS

1 Cut the metal hair comb in half to make two smaller combs.

2 Solder the hair combs to the ends of the half-round wire.

3 Using needle-nosed pliers, curve the half-round wire into a half circle that will fit the head of the wearer. Do this before you solder the embellishments to the curved wire. If you bend the wire after attaching the soldered glass, the bezels may break off of it.

4 Wrap each square and teardrop crystal with the came, starting at the corner of the square crystal or the point of the teardrop crystal. Cut the ends flush. *(photos a–b)*

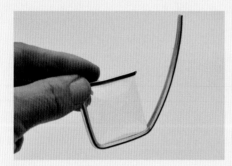

—| A |—

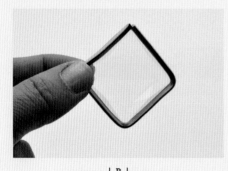

—| B |—

5 Solder a teardrop crystal, a small square crystal, and a large square crystal together as shown. Drop the solder into the joint. Try not to touch the soldering iron to the came, as it will melt just like solder does. Solder the front and back of the piece. Repeat this step to make another identical piece. *(photo c–e)*

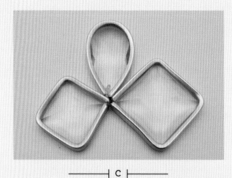

—| C |—

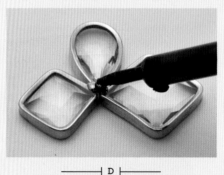

—| D |—

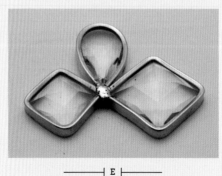

—| E |—

TIP: *Be sure to use glass crystals so they don't melt during soldering.*

6 Solder each 3-crystal piece to the half-round wire. The pieces should fit along the curve and be centered in the front. The bezels of the two large square crystals should meet and touch in the middle

7 Place the last teardrop crystal in the center between the two large square crystals and solder it in place.

8 Thread one of the beads onto each of the headpins. Using round-nose pliers, curl the end of the pins and solder them over each of the joints between the bezels and crystals.

9 Cut the 18-gauge wire into three 3½-inch (8.9 cm) pieces. With the round-nose pliers, make large scrolls on one end of each piece of wire.

10 Solder the scrolled wires in place on the back of the center teardrop crystal near the top.

11 Place the remaining beads on the head pins and secure them by looping end of head pin (see photo on page 81). Hang one from each scroll.

STAMPED SOLDER & COPPER
JEWELRY SET

DIFFICULTY
◆
Easy

YOU WILL NEED

Soft soldering toolkit
Rubber stamp
28-gauge copper sheet metal,
    3 x 5 inch (7.6 x 12.7 cm)
18-gauge wire
2 ear wires
Copper chain

TIPS: *Look for rubber stamps that have deep lines and interesting textures.*

Let the solder run off the edge when you stamp—it'll make a design of its own.

INSTRUCTIONS

1 Make a drawing of the base shape of each piece and then trace each drawing onto the copper sheet metal. Cut the shapes out of the metal.

2 Sand any sharp or rough edges on the metal shapes.

3 Cut a length of wire that's about 2½ inches (6.4 cm) long and use round-nose pliers to shape it into a large loop for one of the pieces. (I used silver-plated wire, but the type you use isn't important.) Make loops in this way for each of the pieces.

4 Solder the wire loop in place on the front of each copper piece.

5 Once the solder has set, add more to make a little mound in the middle of each copper piece.

6 Use the iron tip to keep the mound molten and hot. Then quickly remove the iron tip, press the rubber stamp into the solder, and remove it. The solder should set quickly, and it shouldn't stay on the rubber part of the stamp. If you're not happy with the design, just melt the solder and stamp again!

7 Clean and polish each piece.

8 Attach the ear wires to the earrings and the chain to the pendant.

NOTE: While working with a piece of copper, I accidentally dropped a tool with a rubber handle onto the solder before it was set. The tool created an interesting imprint. The accident (a fortuitous one!) inspired me to start experimenting with texture. I found that if I work on a sheet metal like copper or brass, the solder stays heated long enough for me to stamp it with an object and create an impression. With the exception of a few clear stamps, I found that rubber stamps work surprisingly well and don't melt. Just to be safe, though, I recommend testing a little corner of each of your stamps to make sure it doesn't melt.

TRAVEL JOURNAL PENDANT

DIFFICULTY

✦

Medium

Soft soldering toolkit
Scrapbook papers
2 pieces of glass, each 2 × 3 inches
   (5.1 × 7.6 cm)
Scrapbook embellishments
22-inch (55.9 cm) length of wavy-edge
   copper foil tape, ⁵⁄₁₆ inch (8 mm) wide
Scrapbook cardstock
Scrap of fabric, 2 × 3 inches
   (5.1 × 7.6 cm)
Piece of copper foil sheet, 6 × 12 inches
   (15.2 × 30.5 cm)

Hole punch
Assorted metal charms
2 jump rings, each 5 mm
Black patina
2 standard-size ring shanks
Toggle clasp
2-inch (5.1 cm) length of ribbon
28-inch (71.1 cm) length of tan
   leather cord

NOTE: The cardstock is used for the back cover of each page, for strength and to provide stiffness for the fabric.

INSTRUCTIONS

1 Cut eight 2 × 3-inch (5.1 × 7.6 cm) pieces of scrapbook paper to create four sets of papers. Pair up the pieces and glue the front and back sides of them together. You should end up with a total of four glued sets of papers.

2 Cut two more 2 × 3-inch (5.1 × 7.6 cm) pieces of scrapbook paper for use as the front cover (inside and outside) of the journal. Decorate the paper any way you like using the scrapbook embellishments. Then glue the pieces together and place them between the pieces of glass. Wrap the edges of the glass with the copper foil tape.

3 Cut one 2 × 3-inch (5.1 × 7.6 cm) piece of cardstock and one 2 × 3-inch (5.1 × 7.6 cm) piece of fabric. Sandwich them together and wrap them with copper foil tape.

4 Cut six 2 × 1½-inch (5.1 × 3.8 cm) pieces from the copper foil sheet. Fold the pieces in half lengthwise so that each one measures 2 × ¾ inches (5.1 × 1.9 cm). Use the hole punch to make two holes at the top edge of each piece. Use the first one as a guide for the remaining five by placing it on top of each subsequent piece you punch. (photo a)

5 At the top of each page and the front cover and back covers, apply the punched copper foil tape. Overlap each edge on either side by ¼ inch (6 mm) or so. The edge of the glass, paper, and fabric should be just under the bottoms of the holes you punched. (photos b–c)

6 Apply solder to all of the copper-foiled surfaces. Be careful not to get flux or solder on the paper. You only need a thin coat of solder.

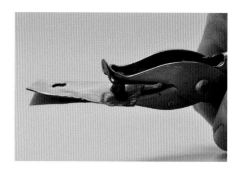

| A |

| B |

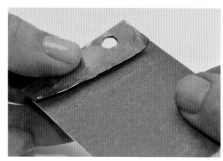

| C |

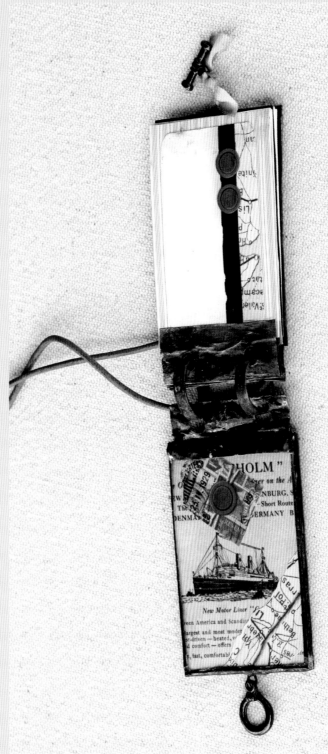

7 Tack-solder the decorative charms to the front cover.

8 Solder a small jump ring to each of the covers along the bottom edge, matching them up to be centered.

9 Apply black patina to the soldered areas, the charms, and the ring shanks. Use a small amount of liquid patina so you don't get the paper wet.

10 Attach the loop part of the toggle clasp to the jump ring on the front cover. Attach the bar part of the clasp to the jump ring on the back cover using the ribbon.

11 Thread the ring shanks through the holes at the tops of the pages and covers. Overlap the ends to close the rings. Overlap them more to make them smaller or less to make them larger.

12 Thread the leather cord through the ring shanks and make an adjustable sliding knot to finish the necklace.

REVERSIBLE HALLOWEEN
& THANKSGIVING ORNAMENT

DIFFICULTY

✦

Easy

Soft soldering toolkit
Halloween- and Thanksgiving-themed
  decorative papers
2 pieces of glass, each 1¼ × 3 inches
  (3.2 × 7.6 cm)
2 pieces of glass, each 2 × 3 inches
  (5.1 × 7.6 cm)

Copper foil tape, 5/16-inch (8 mm) wide
4 large jump rings
2 decorative wire swirls
Orange acrylic paint
Decorative ribbon, 8 inches
  (20.3 cm) long

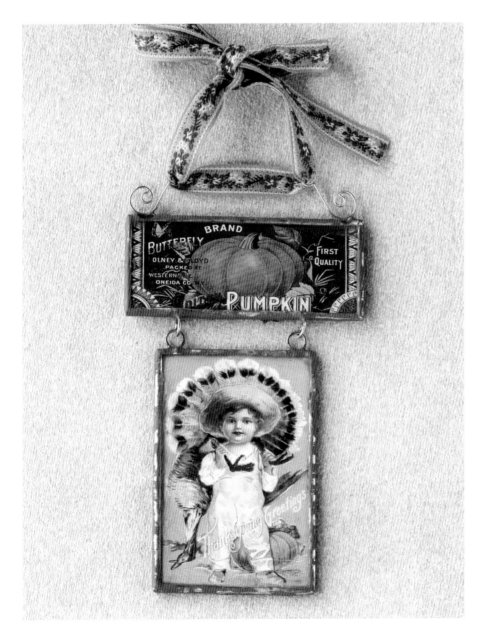

INSTRUCTIONS

1 Cut pieces of the Halloween and Thanksgiving papers that will fit between the sheets of glass. You'll need papers in two sizes to match the glass pieces: 1¼ × 3 inches (3.2 × 7.6 cm) and 2 × 3 inches (5.1 × 7.6 cm). Place the Halloween and Thanksgiving designs back-to-back and sandwich them between the pieces of glass.

2 Wrap the edges of the glass pieces with copper foil tape and solder them.

3 Line the pieces up and mark spots for the jump rings. The jump rings will connect the two glass pieces.

4 Solder two jump rings in place on the lower end of the top piece. Solder the decorative swirls at an angle near each top corner of the top piece for the ribbon.

5 Connect 2 more jump rings to the ones you soldered onto the top piece. Carefully solder them onto the bottom piece so that the closure is soldered shut.

6 Clean off the flux and sand the soldered edges to give them some "tooth." *(see photos)*

7 Paint the soldered edges with the acrylic paint.

8 When the paint is dry, lightly sand it in some areas to give the piece a rustic vintage look.

9 Tie the ribbon onto the swirls.

STAMPED SOLDER PENDANT

DIFFICULTY
✦
Easy

Soft soldering toolkit
Piece of 28-gauge copper sheet metal
Rubber stamp
2 jump rings

Black patina
Thin paintbrush
Premade Greek leather necklace

TIP: The copper sheet is thin enough to be cut with sturdy scissors. The scissors will give you more control than tin snips or metal shears.

INSTRUCTIONS

1 Cut the copper sheet to the size you'd like your pendant to be. Smooth the edges as necessary with a hand file.

2 Melt a thick layer of solder over the whole front side, covering all of the copper. Keep the solder hot and molten.

3 Remove the iron and very quickly stamp the solder with the rubber stamp.

4 Carefully solder the jump ring onto the back of the pendant, making sure you don't heat it up too much. If you do, the front design could melt.

5 Clean the pendant by rinsing it with water.

6 Use a tiny paintbrush to paint black patina into the recessed areas of the solder.

7 Attach the pendant to the pre-made necklace using the second jump ring.

STARFISH NECKLACE

DIFFICULTY
◆
Medium

Soft soldering toolkit

Starfish metal embellishments

Assorted mini shells

2 watch crystals, similar in shape and size

Metal bead with large hole

Copper, silver-backed foil tape, 1/2-inch (1.3 cm) wide, approximately 4 inches (10.2 cm) long

Rotary tool with a sanding bit

7-inch (17.8 cm) length of beading wire

24 mini shell beads

2 silver crimp tubes

24-inch length of Greek leather

TIPS: *You can easily find watch crystals on eBay. Make sure the ones you use are made from real crystal. Plastic ones will melt when you solder them.*

Use silver-backed foil tape when you work with thick, clear glass, because it will show through. Silver usually looks better than copper.

INSTRUCTIONS

1 Place the mini shells and the starfish embellishment between the two watch crystals.

2 Wrap the edges of the crystals with copper foil tape. There will be a little gap between the edges of the two crystals, and the embellishments and shells will fall through it while you're wrapping, if you aren't careful. Solder the edges of the crystals.

3 Solder the metal bead to the top of pendant.

4 Using the rotary tool with the sanding bit, smooth the soldered surfaces to create a brushed metal appearance.

5 String the beading wire through the large soldered bead at the top of the pendant. String the mini shell beads in a pattern so crystal ornament is in the center.

6 Finish the ends of the wire with small loops and crimp tubes.

7 Cut the length of Greek leather in half and thread it through the ends of the loops on the shell strand. Secure it with knots.

8 Tie a sliding knot to finish the necklace and make it adjustable.

STARFISH BRACELET

DIFFICULTY

✦

Easy

Three metal starfish embellishments
Soft soldering toolkit
2½-inch (6.5 cm) length of 21-gauge, silver-plated wire
12 pearl beads

8-inch (20 cm) length of bead wire
2 crimp tubes
9 seed beads
Lobster clasp
Small jump ring

INSTRUCTIONS

1 Fill in the back of each starfish with solder.

2 Set the first starfish facedown and solder the 2½-inch silver-plated wire to the back of it, leaving an equal amount of loose wire on either side.

3 String a pearl bead onto the wire on each side of starfish and solder the next starfish on the other side of bead. Repeat on the other side with another pearl bead and the final starfish.

4 String one pearl on each side of the last starfish and make a wrapped loop on each end of the wire.

5 Make a loop on one end of the bead wire and secure it with a crimp tube. String the pearl beads and seed beads onto the wire until it's the right length. (I like to make this a snug bracelet to keep it from twisting on my wrist.) Finish the other end with another loop and crimp tube.

6 Attach a lobster clasp to the end with a small jump ring. To close the bracelet, connect the lobster clasp to the wrapped loop at the end of the starfish chain.

TIP: *Check out the scrapbooking aisles of the craft stores—they often have interesting metal embellishments you can use to solder with.*

WEDDING FRAME

YOU WILL NEED

Soft soldering toolkit
Photograph
Two pieces of rectangular glass, each
    3½ × 5 inches (8.9 × 12.7 cm)
Foil tape, ⁵/₁₆-inch (8 mm) wide,
    18 inches (45.7 cm) long

36-inch (91.4 cm) length of heavy-gauge
    twisted wire
Heavy-duty pliers
Five jump rings
Seven pearl-and-crystal bead charms

INSTRUCTIONS

1 Place the photo between the pieces of glass and wrap it with foil tape.

2 Solder the edges.

3 Cut two 12-inch (30.5 cm) pieces of the heavy wire and bend them in half using round-nose pliers.

4 Using strong pliers, curl up the ends of each wire. Hold the wires together to make sure both pieces are the same size.

5 Bend the curled pieces in slightly to form a V shape. Solder in place at the top part of the V. These will become the stand for the frame.

6 Hang a bead charm from the curled loop on the end of the wire.

7 Cut the last 12-inch (30.5 cm) section of wire in half.

8 Bend the wires into a scroll with a large spiral on one end and a small spiral on the other. Match them up so that they're the same size.

9 Solder the scrolls to the top corners of the frame so that the smaller spirals meet in the middle.

10 Solder a jump ring where the two spirals meet. Attach three bead charms to the jump ring.

WIRE COIL CUFF

DIFFICULTY

♦

Easy

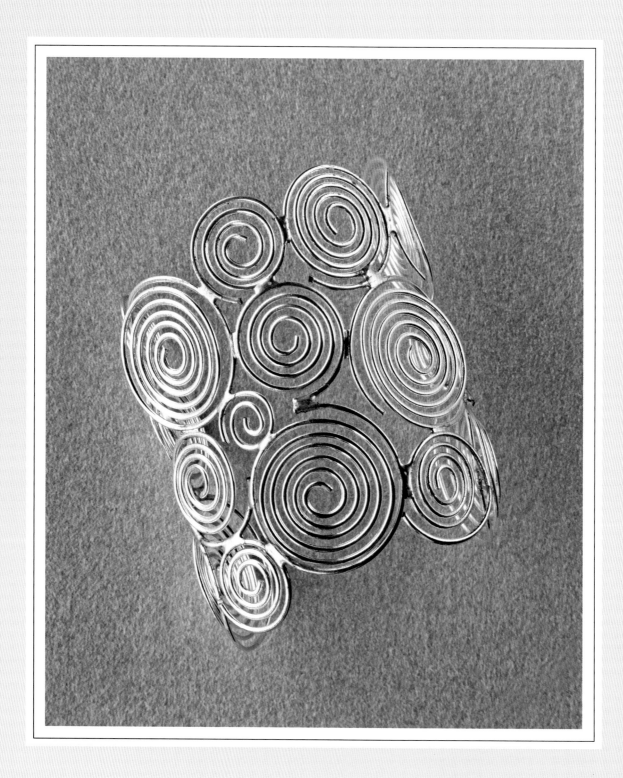

Soft soldering toolkit
20 preformed metal wire wrap coils,
    each 20 mm
Bracelet-bending pliers

TIPS: *You can purchase preformed metal coils at your local craft store. They're available in many different sizes. If you prefer, you can make them by hand.*

If any of the joints comes apart while you're forming the bracelet, you can easily re-solder it.

INSTRUCTIONS

1 Cut six of the coils down to $1/2$ inch (1 cm). Cut five of them to $1/4$ inch (5 mm). Cut two or three more coils to 1-inch (2.5 cm as needed to fill in space.

2 Set the coils out in a random pattern, matching them up so that the ends are on an inner section of the cuff, and each coil is touching the other coils in two or three spots that you can solder.

3 Solder the coils together where they touch, starting on one end and working across.

4 Use a pair of bracelet-bending pliers to form the soldered coils into a curved cuff. Finish as described on pages 23 and 24, and rinse, sand, and polish as necessary.

SECRET GARDEN DIORAMA

YOU WILL NEED

Soft soldering toolkit
2 pieces of glass, each 5 × 3½ inches (12.7 × 8.9 cm)
2 pieces of glass, each 2½ × 3½ inches (6.4 × 8.9 cm)
Silvered copper foil tape, ½-inch (1.3 cm) wide
Popsicle stick
Round-nose pliers
4 decorative metal corners
Photos of people and an animal
Decorative paper butterflies
Decorative paper and ribbons

Metal crown charm
Heavy piece of cardboard, 2½ × 5 inches (6.4 × 12.7 cm)
Craft glue
Dried craft moss
26-inch (66 cm) length of 18-mm silver-plated wire
Plain white paper
Sheet of standard aluminum foil tape, 4 inches (10.2 cm) wide
Decorative paper flowers
Decorative head pin
Large teardrop bead

TIP: *To get the odd-sized pieces of glass listed here, you may need to buy bigger pieces. Try 5 × 7 inches (12.7 × 17.8 cm). Then cut them down. Check your craft store for inexpensive picture frames and pre-cut glass pieces that will fit together so that you don't have to cut anything.*

INSTRUCTIONS

1 Lay the glass pieces out as shown, in the order in which they'll be assembled. *(photo a)*

2 Cut an 8-inch (20.3 cm) length of the silvered copper foil tape.

3 Place the first two sides of glass together—one of the short ends on a 5 × 3½-inch (12.7 × 8.9 cm) piece and a long edge of a smaller piece. Place them at the center of the foil tape, allowing for space between the two pieces of glass so the tape has room to bend and form a corner. *(photo b)*

4 Wrap the foil tape all the way around so that it covers both sides of the glass where the pieces meet, overlapping anywhere on the side that will be the inside of the box. Smooth the tape in place. *(photo c)*

5 Attach the next piece of glass the same way, followed by the final piece.

6 To form the fourth corner, bring the first piece and the last piece together and wrap the foil around it the same way you did with the other pieces to create the box. Press the foil tape with a Popsicle stick to crease the insides of the corners.

7 If you can see the back of the tape that's inside the box from the outside, then trim the tape with a craft knife.

8 Solder each of the taped corners inside and out, being careful to maintain the rectangular shape of the box.

9 Apply tape to the top and bottom edges of the glass and solder it.

A

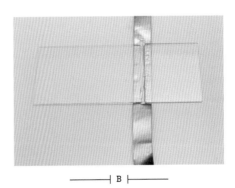

B

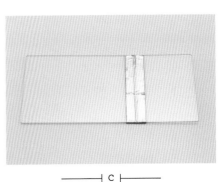

C

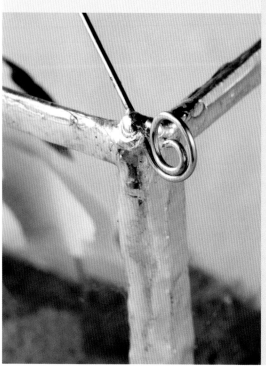

10 Using a pair of round-nose pliers, bend the decorative metal corners so they fit around the corners of the box. Solder them in place, making sure the box sits level.

11 Cut the people out of the photo. Cut wings from the butterfly paper.

12 Create a cone hat for the animal from decorative paper. Decorate the animal with embellishments or ribbons.

13 Put the crown charm and a bow (or the embellishments of your choice) on the people.

14 Trim the cardboard so that it fits snugly inside the bottom of the box. Make sure it fits and remove it from the box. Glue the moss in place on the cardboard. Work on steps 15 through 19 out of the box.

15 Cut 5-inch (12.7 cm) pieces of the silver-plated wire for the people and the animal. Form the wire into U shapes and glue them to the back of the people and the animal, leaving ½ to 1 inch (1.3 to 5.1 cm) of length at the bottom to attach them to the box. Glue matching paper over the exposed wire to cover it.

16 Crease the wings in the center so that they have dimension, and then glue them to the back of the people.

17 Poke the ends of the wire through the cardboard, bending them in opposite directions

so that they lie flat against the underside. Adjust the people and the animal as needed by bending the wire.

18 Secure the ends of the wire in place by covering the entire bottom side of the cardboard with aluminum foil tape to give it strength and a nice silver finish.

19 Glue the small paper flowers to the moss.

20 Place the finished diorama inside the soldered box from the top. Press it down, securing it into the bottom of the box. Make sure it fits snugly and can't move around in the box.

21 Cut two lengths of wire, each about 8 inches (20.3 cm) long. Curl the ends of both wires and curve them to create a slight arch. Solder them in place at the top corners of the box, crossing them in the middle.

22 Thread the head pin through the bead, bend the end of the pin at an angle, and trim it.

23 Solder the head pin to the top of the wires where they cross.

BEDAZZLED BELT BUCKLE

DIFFICULTY
✦
Medium

Soft soldering toolkit
Piece of glass, 1½ inches (3.8 cm) square
  and 4 mm thick
Solid sheet of copper foil tape,
  2 × 2 inches (5.1 x 5.1 cm)

18-gauge heavy silver-plated wire,
  3 inches (7.6 cm) long
Decorative punch
10-inch (25.4 cm) length of glass
  rhinestone chain

TIPS: *Don't use plastic rhinestones with this project because they'll melt and lose their shine.*

If no punch is available, draw your own design and cut it out using a craft knife.

INSTRUCTIONS

1 Press the glass onto the copper foil tape to create an outline of the piece. Then remove glass *(photo a)*

2 Punch the decorative element out of the foil tape using the punch. You'll need to apply heavy pressure to cut through the foil. I placed the project on the floor and used my foot to gently squeeze the punch.

3 Peel the backing off the copper foil and place the glass on the outline you created earlier.

4 Wrap the copper foil around the sides to the back of the glass, mitering the foil by folding and tucking it in at the corners.

5 Smooth and burnish the foil in place.

6 Trim the foil as needed if it's showing through the punched design.

7 Solder over all of the foil.

8 Cut two pieces of the wire, one to fit each side of the glass with about an extra ¼ inch (6 mm) left over on each end.

9 Bend the ends of each wire into right angles that face the same direction.

10 Solder the ends in place so that the wires are parallel to the two opposite sides of the buckle, leaving a little bit of space between the wire and edge so that you can pass a belt through the opening.

11 Cut four lengths of the rhinestone chain that will fit each edge of the buckle. Slide the chains together to close the gaps between the stones.

12 Holding the chain with pliers, solder all along the back side of the metal on each chain to keep the rhinestones snug together. This will make it easier for you to solder them onto the buckle.

13 Tack-solder the rhinestone chains to the edges of the buckle.

⊢ A ⊣

BIRDCAGE PENDANT

DIFFICULTY

Medium

YOU WILL NEED

Soft soldering toolkit
Decorative paper
2 pieces of identically shaped glass,
    each about 1½ × 2 inches
    (3.8 × 5.1 cm)
Small image of a bird
Craft glue
Foil crown
8-inch (20.3 cm) length of 16-gauge
    silver-plated wire
¼-inch (6 mm) copper foil tape, 6 inches
    (15.2 cm) long

2 large jump rings
Large pear-shaped pearl bead
Head pin
Silver bead cap
Silver daisy spacer
3 decorative links, each 1 cm
19-inch (48.3 cm) length of link chain
6 medium jump rings
Lobster clasp
Small jump ring

INSTRUCTIONS

1 Cut a piece of the decorative background paper that will fit between the pieces of glass.

2 Cut out the bird image and glue it to the decorative paper. Glue the foil crown onto the bird's head.

3 Sandwich the paper between the pieces of glass and wrap it with ¼-inch (6 mm) copper foil tape. Solder all of the edges.

4 To make the wire cage element, start with the center wire. Cut a length that will go from the top of the soldered glass shape to the bottom with an extra ¾ inch (1.9 cm) left over.

5 Curve the wire so that it bulges out and away from the glass. Make a curled loop at each end of the wire with round-nose pliers. Leave enough space in the loop for a jump ring later.

6 Make the other wires in the necessary lengths to cover the front.

7 Tack-solder each wire in place starting in the middle and working out to the sides.

(continued on page 109)

8 Create the pearl bead charm by stringing the daisy spacer, the pearl, and the bead cap onto the head pin. Make a wrapped loop at the end of the pin (see page 107). Hang the pearl bead charm from the loop at the bottom of the center cage bar.

9 Cut two 6-inch (15.2 cm) lengths, two 1½-inch (3.8 cm) lengths, and two 2-inch (5.1 cm) lengths of chain.

10 Attach a 1½-inch (3.8 cm) length of chain and a 2-inch (5.1 cm) length of chain to the center decorative link using one medium jump ring. Repeat this step so that you have a total of four chains attached to the link with two medium jump rings.

11 Connect the other ends of the four chains to the remaining two decorative links using two more medium jump rings.

12 Attach a 6-inch (15.2 cm) chain with a medium jump ring to the other side of each decorative link.

13 Finish the chain off with a large jump ring and a lobster clasp.

14 Attach a large jump ring to the center decorative link, and solder the open side of the jump ring to the top of the birdcage charm.

BOTTLE BRACELET

Soft soldering toolkit
Wine bottle, standard size
Bottle cutter
Butane hand torch
12-inch (30.5 cm) length of silver-
    backed copper foil tape, ½-inch
    (2.5 cm) wide

6 jump rings, each 3 mm
3 lengths of clear stretch cord, each
    3 inches (7.6 cm) long
36 oval metal beads, each 6 mm

TIPS: *Choose a thick bottle
for sturdiness.*

*Broken pieces can be
fashioned into charms
and other elements.*

*Find bottles with unique
designs already painted
on them.*

*Use silver-backed copper foil
tape when working on glass
you can see through.*

INSTRUCTIONS

1 Use the bottle cutter's guide to adjust where the first cut will be. *(photo a)*

2 Pressing firmly, spin the bottle over the cutting blade to complete a circle around the bottle. You'll hear the glass making a crunching sound as it cuts, and you'll see a score line. *(photo b)*

3 Use the hand torch to heat along the score line of the bottle until you hear a slight cracking noise. If you don't have a hand torch, use a candle flame. *(photo c)*

4 Apply ice to the score line. This will cause the crack to continue until the bottle comes apart. You may need to gently tap the bottle on a table to help break it apart. *(photos d–e)*

5 Repeat the process a second time to cut out the section of the bottle you want to work with.

6 Use grozing pliers to "bite" off any jagged edges of glass. Smooth off any rough or uneven edges with sandpaper. (Be sure to wear goggles and gloves for protection.) *(photos f–g)*

7 Starting at one end, wrap the copper foil tape around the inner and outer edges of your glass piece as shown. *(photo h)*

8 On the inside edge of the glass, use sharp scissors to cut small slits in the foil. This makes it easier to fold the foil in. *(photo i)*

9 Bend the notched side of tape in over the edge of the glass and smooth it into place. *(photo j)*

——| A |——

——| B |——

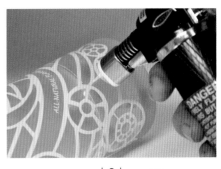

——| C |——

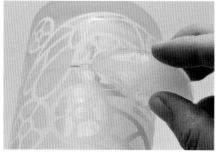

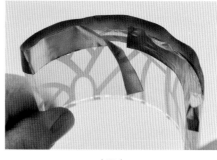

| D |

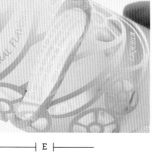

| E |

| G |

| I |

TIP: *Cutting the bottle was trickier than I expected it to be. Keep in mind that the glass can crack in other places besides the score line when you're heating, icing, or soldering it.*

10 Fold the outer side of the foil down over the notched foil. Smooth out all bumps.

11 Repeat the foiling process on the other edge.

12 Cover each end of the glass piece with foil to complete the frame. *(photo k)*

13 Solder all of the foiled edges.

14 Solder three jump rings to each end of the bracelet, making sure they are evenly spaced.

15 String the beads onto the stretch cord. Attach the pieces of cord to the jump rings, and tie knots to secure them.

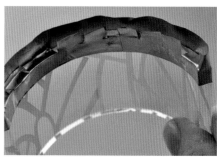

| J |

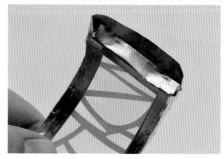

| K |

HARD
SOLDERING

BEZELED ROSE QUARTZ

DIFFICULTY
✦
Easy

Hard soldering toolkit
Rose quartz stone, approximately
 ½-inch (1.5 cm) diameter
Sheet of paper
Fine silver bezel strip, ³⁄₂₀ × 3 inches
 (4 mm × 7.6 cm)

Sterling silver solder chip, easy
Sterling silver 5 mm (0.2-inch)
 jump ring
Industrial-strength craft glue
 (e.g., E6000)

| A |

INSTRUCTIONS

1 Wrap a piece of paper around the quartz stone to create a pattern for the bezel strip. *(photo a)*

2 Bend the bezel strip around the outer edge of the stone, forming it to the stone's contours.

3 Trim the ends of the bezel strip straight across so that they meet flush, without overlapping.

4 Remove the formed bezel from the stone and place it on your soldering surface, making sure the ends are touching with no gap. *(photo b)*

5 Coat the joint with flux. *(photo c)*

6 Heat the joint with your torch until the paste flux looks clear and becomes tacky.

| B |

| C |

7 Place a small piece of easy-solder chip with flux on it on the joint.

8 Heat the joint with the torch until it's filled and the solder is smooth. *(photo d–e)*

9 File the joint to smooth out any lumps.

10 Melt solder onto the jump ring where it closes so that it's soldered shut. Flux the bezel where the jump ring will go. Remelt the jump ring solder and let it flow onto the bezel to solder it in place. *(photo f–g)*

11 Place the finished bezel in the pickle until clean.

12 Clean and polish the project.

13 Dab a small amount of industrial-strength craft glue around the whole inside of the bezel, and place the stone inside. Smooth down the bezel's edges using a burnisher so that it fits tightly around the stone.

14 String the bezeled stone onto the necklace or chain of your choice.

D

E

F

G

STERLING SILVER
SWEAT SOLDER RING

DIFFICULTY

✦

Challenging

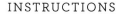

Hard soldering toolkit
Piece of 30-gauge sterling silver sheet
metal, 2 x 3 inches (5.2 x 7.6 cm)
Ring mandrel or cylindrical item
Piece of cardboard

Pounding anvil
Texture hammer
28-gauge fine sterling silver wire, at least
26 inches (66 cm) long

TIPS: *If you're new to soldering, this project may seem a bit intimidating. This ring involves multiple techniques (creating a ring, making a textured base, etc.), but any of the techniques can also be used on its own.*

If the joint on your ring isn't perfect, don't worry—it can easily be hidden under the center detail.

INSTRUCTIONS

1 Measure your finger with a piece of paper to create a template for your ring. *(photo a)*

2 Cut a strip of the sterling silver sheet metal the size of your template.

3 Using a ring mandrel or some other cylindrical item, form the sheet metal strip into a ring.

4 Make sure the ends of the sheet metal meet up exactly. Trim or file it if necessary to get a tight fit. *(photo b)*

5 Solder the inside of the seam. *(photo c)*

6 Cut a 1 x ½-inch (2.5 x 1.3 cm) rectangle from the sheet metal. Round the corners.

7 Place the rounded rectangle of silver on a piece of corrugated cardboard on top of the pounding anvil. Use a texture hammer to create a textured finish on the metal. *(photos d–e)*

8 Cut four 3-inch (7.6 cm) lengths of the sterling silver wire.

⊢ A ⊢

⊢ B ⊢

⊢ C ⊢

———| D |———

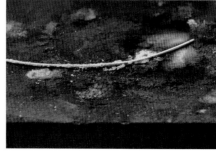

———| F |———

———| J |———

———| E |———

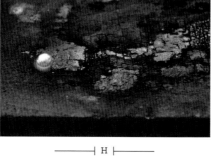

———| G |———

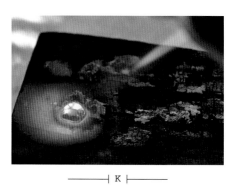

———| K |———

9 Flux one piece of 3-inch (7.6 cm) wire. Starting at one end, heat it until it forms a ball that crawls up the wire. Be careful, as the molten ball can easily slip off your work surface. Use the torch to guide it. Repeat this process with each of the other 3-inch (7.6 cm) wires. *(photos f–h)*

———| H |———

10 Place the four balls together and flux them. *(photo i)*

11 Heat the four balls together, running the flame in a circular motion to create one large ball. Pickle and wash the resulting large ball. *(photos j–k)*

12 Use a hammer with a flat head to pound the ball into a flat disc. *(photo l)*

———| I |———

13 Cut two 4-inch (10.2 cm) pieces of the sterling silver wire.

14 Make two balls and melt them together as you did in step 9. Pickle, wash, and flatten this ball.

15 Cut a 6-inch (15.2 cm) piece of wire, flux it, and melt into a ball. Pickle and clean it; don't smash it.

16 Place the ball on your work surface with the flatter side up. *(photo m)*

17 Flux the ball, and then melt a small amount of medium solder onto it. *(photos n–o)*

18 Flux the second flat disc and place the ball on the center of it. Heat it in a circular motion, using the solder pick if needed to hold the ball in place. When the ball sinks down, it has been soldered. *(photos p–q)*

— L —

— M —

— N —

— O —

19 Flux the largest disc in the center and place a piece of easy solder on it. Heat it and let it start to melt.

20 Flux the back of the medium-sized disc. Place it in the center of the large disc.

21 Heat the piece all around in a circular motion until the medium disc sinks into place.

22 Flux and melt a piece of easy solder in the center of the textured rectangle.

23 Flux the back of the disc stack and place it in the center over the solder spot.

24 Heat the piece in a circular motion until the whole stack sinks into place. *(photo r)*

25 Flux and melt some easy solder on the ring shank where the seam is.

26 Place the stacked piece face-down on your work surface. Apply pressure with your fingers to press it into the charcoal block.

27 Heat it in a circular motion. Move the flame down, in between where the ring and the back join to get extra heat to the solder. When the ring sinks in place, remove the flame. *(photo s)*

28 Pickle and rinse the ring. Shine it with a brass-bristled brush to finish.

— P —

— Q —

— R —

— S —

STERLING FILIGREE EARRINGS

DIFFICULTY
◆
Easy

YOU WILL NEED

Hard soldering toolkit
2 pieces of 28-gauge sterling silver sheet metal, each 1¼ × 1 inch (3.2 × 2.5 cm)

12-inch (30.5 cm) length of twisted 18-gauge sterling silver wire (filigree)
Sterling silver solder, medium and easy
2 sterling silver ear wires

———┤ A ├———

INSTRUCTIONS

1 Round the corners slightly on each piece of sheet metal using a file or a rotary tool.

2 Cut two 5½-inch (14 cm) pieces and two ½-inch (1.3 cm) pieces from the sterling silver wire.

3 Create loops at both ends of the 5½-inch (14 cm) pieces, making one loop bigger than the other. Bend the wire so that both loops open in the same direction.

4 Bend the ½-inch (1.3 cm) pieces of wire into smaller loops. Solder them to the longer pieces with medium solder.

5 Paint the front sides of each piece of the sheet metal with flux. *(photo a)*

6 Set the filigree on the sheet metal. Make sure it's lying flat on the surface with no gaps. If there are gaps, adjust the wire or the base of sheet metal—try bending it by hand or pounding it lightly with a jewelry mallet. *(photo b)*

7 Apply the flux to the filigree where you want to tack-solder it in place. You only need to solder a few spots, not the entire piece.

8 Put a small piece of easy silver solder in the spots where you'll be soldering the joint. You can do this all at once, but the solder may float or migrate out of place. If this happens, use the solder pick to push it back into place. You can do each spot individually, if you prefer. *(photo c)*

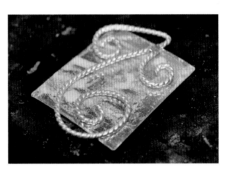

———┤ B ├———

9 Heat all of the metal by passing the torch in a circular motion around the whole piece, pausing and concentrating on the solder a second longer, but moving in a continuous motion.

———┤ C ├———

10 Once the solder is molten and flows into place, remove the heat. If the filigree moves or floats a little, use the pick to hold it down while you heat it, until you remove the flame and the solder sets.

11 Pickle, clean, and polish the earrings.

12 Add the sterling ear wires.

TIP: *You can't fill in gaps with solder, so when you create a piece of jewelry, things need to sit just right. To achieve this, use a solder pick. You can use it to hold part of a project flat on the surface you're soldering it to if there are gaps between the two parts.*

FLOWER-AND-VINE NECKLACE

DIFFICULTY

✦

Easy

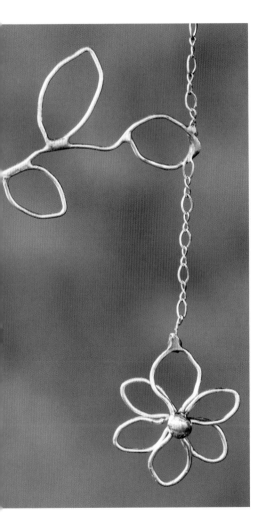

Hard soldering toolkit
16-inch (40.6 cm) length of 20-gauge
    fine sterling silver wire
Sterling silver solder, medium and easy

18-inch (45.7 cm) length of sterling silver
    chain
2 sterling silver jump rings
Sterling lobster clasp

INSTRUCTIONS

1 Cut a 3½-inch (8.9 cm) piece of wire. Use the round-nose pliers to curl one end of the wire into a small spiral and the other end into a leaf shape. Use medium solder to solder the leaf closed at the end.

2 Cut two 1½-inch (3.8 cm) pieces of wire and two 1-inch (2.5 cm) pieces of wire. Shape each piece by bending it in half and rounding out the sides with the pliers to create a leaf shape.

3 Use medium solder to solder each leaf closed.

4 Use easy solder to solder the leaves to the 3½-inch (8.9 cm) piece of wire.

5 To make the flower, cut three 2-inch (5.1 cm) pieces of wire and form each of them into a circle. Use medium solder to solder the circles closed. Use needle-nose pliers to pinch each circle together in the middle, to form two petals. Repeat for two other circles.

6 Set the formed petals on top of each other and use medium solder to solder them in place in a flower shape.

7 Cut a 1½-inch (3.8 cm) piece of wire and melt it until it forms a ball. The ball will serve as the center of the flower.

8 Melt a piece of easy solder onto the bottom of the ball. Place the ball in the center of the flower. Sweat-solder the ball in place.

9 Cut the silver chain into two pieces. One side should be 1 inch (2.5 cm) longer than the other. Melt the longer chain end onto the top end of one flower petal using easy solder.

10 About 1 inch (2.5 cm) up from the top of the flower, solder the chain to the leaf at the end of the vine using easy solder.

11 Attach by threading the other chain onto the spiral end of the vine. (Do not solder in place, as the end needs to move freely.)

12 Finish the other end of the chain with a lobster clasp and jump rings.

STERLING BEZEL PENDANT

DIFFICULTY
◆
Challenging

Hard soldering toolkit
30-gauge fine silver bezel strip,
   4 mm wide
Large oval stone with flat back,
   approximately 1 x ⁴/₅ inches
   (2.5 x 2 cm)
Large piece of sandpaper
Piece of 28-gauge sterling silver sheet
   metal, 1 × 1½ inches (2.5 x 3.8 cm)
   for backing

Silver solder, medium and easy
Jewelry cement
Drill with very small bit
Bead reamer
18-inch (45.7 cm) length of sterling silver
   chain
3 jump rings
Lobster clasp

TIP: *Since solder will not fill in a gap, things need to sit just right. Using a pick is a handy way to "cheat." If you can't get your project to lay flat to the surface you're soldering it to, you can use a pick to force it to lay flat by pressing it in place.*

INSTRUCTIONS

1 Measure the stone with a strip of paper to determine the bezel length. The measurement needs to be exact—make sure the ends of the paper meet without any gaps or overlap.

2 Use the paper pattern to cut the bezel strip. Cut the end flush and sand it if necessary to smooth any uneven edges.

3 Wrap the bezel strip around the stone to make sure that it fits evenly and the ends meet.

4 Take the bezel strip off of the stone and solder the joint with medium solder.

5 Wrap the bezel around the stone again to make sure it still fits. If it's too small, start over. If it's too big, cut it apart, trim it so that it fits, and solder it again. Once the stone fits, remove it from the bezel.

6 Sand or smooth the joint, if necessary.

7 Use a large piece of sandpaper to sand the bottom edge of the bezel (moving it in a figure-eight pattern) to ensure that the bezel rests flat on the backing piece. *(photo a)*

8 Set the bezel in place on the sheet metal backing to make sure it sits flat without any gaps, adjusting it as necessary. (A few small gaps are okay as long as they aren't noticeable.)

9 Make sure the stone still fits in the bezel.

10 Flux the front side of the sheet metal and the entire bezel strip. Place the strip where you want to solder it. *(photo b)*

11 Place pieces of easy solder around the inside of the bezel right next to the joint, spacing them about ⅛ inch (3 mm) apart. You should put the solder on the inside so that it can't be seen from the outside. *(photo c)*

| A |

| B |

| C |

D

E

F

12 Heat the entire piece in a circular motion. Make sure to keep the backing well heated.

13 Once the solder begins to melt, make sure the bezel is still in the right place. Use a solder pick to move it, if necessary. You can also use the solder pick to hold the bezel down flat as the solder sets. A few stray pieces of solder will affect how the stone fits in the bezel. I used a slightly curved stone, which ensured that the bumps of solder around the edges didn't get in the way. If the bumps do get in the way, sand them off or re-melt the solder to smooth it out. *(photo d)*

14 Remove the heat, pickle the piece, and clean and polish it. I used a light-grit sandpaper to give the silver a brushed appearance. If you don't sand the piece, it will have more of a shine. Trim back the piece by rounding the corners. If the bezel moves out of place and is off-center, you can trim around the outside at this time to make it more even on all edges.

HINT: *Creating a brushed finish is an easy way to camouflage scratches or imperfections. It also gives the piece an antique or rustic feel.*

G

15 Drill a hole in the upper corners of the pendant with the hand drill. Use the bead reamer to make the holes larger and to sand them smooth. *(photos e–f)*

16 Place the stone in the bezel. Put a dab of jewelry cement in the base to hold the stone in place.

17 Burnish the upper edges of the bezel strip to fit the curve of the stone. *(photo g)*

18 Thread the chain through the holes and finish the pendant using the jump rings and lobster clasp.

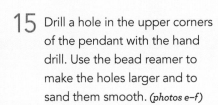

ACKNOWLEDGMENTS

———— ✦ ————

Thank you to all the creative people who share their ideas and inspire me. Thank you to my family and friends for encouraging me. Lastly thank you to Jessica my biggest fan & sister queen for life.

ABOUT THE AUTHOR

———— ✦ ————

Lisa Bluhm is a passionate, creative mixed media artist. In addition to her savvy soldering skills she has created her own soldered art product line called Simply Swank and shares her inspiration in classes and at conventions all across the United States. When she is not in the studio, you can find her teaching dance fitness classes or representing her state as the 2011 Mrs. Washington America pageant queen. She lives in the beautiful Pacific Northwest with her husband, Allen, sons Cameron and Jordan, and her pet Schnoodle. Visit LisaBluhm.net to find Lisa's projects and teaching schedule and for information about sourcing materials for your soldering projects.

INDEX